U.S. NAVAL I

Chro

For nearly a century and a half since a group of concerned naval officers gathered to provide a forum for the exchange of constructive ideas, the U.S. Naval Institute has been a unique source of information relevant to the nation's sea services. Through the open forum provided by *Proceedings* and *Naval History* magazines, Naval Institute Press (the book-publishing arm of the institute), a robust Oral History program, and more recent immersion in various cyber activities (including the *Naval Institute Blog* and *Naval Institute News*), USNI has built a vast assemblage of intellectual content that has long supported the Navy, Marine Corps, and Coast Guard as well as the nation as a whole.

Recognizing the potential value of this exceptional collection, USNI has embarked on a number of new platforms to reintroduce readers to significant portions of this virtual treasure trove. The U.S. Naval Institute Chronicles series focuses on the relevance of history by resurrecting appropriate selections that are built around various themes, such as battles, personalities, and service components. Available in both paper and eBook versions, these carefully selected volumes help readers navigate through this intellectual labyrinth by providing some of the best contributions that have provided unique perspectives and helped shape naval thinking over the many decades since the institute's founding in 1873.

THE U.S. NAVAL INSTITUTE ON
WOMEN IN THE NAVY
THE HISTORY

THE U.S. NAVAL INSTITUTE ON
WOMEN IN THE NAVY
THE HISTORY

THOMAS J. CUTLER
SERIES EDITOR

Naval Institute Press
Annapolis, Maryland

Naval Institute Press
291 Wood Road
Annapolis, MD 21402

Library of Congress Cataloging-in-Publication Data
United States Naval Institute.
  The U.S. Naval Institute on women in the Navy : the history.
     pages cm. — (U.S. Naval Institute chronicles)
  Includes index.
  ISBN 978-1-61251-984-5 (alk. paper) — ISBN 978-1-61251-985-2 (ebook)
1. United States. Navy—Women—History. 2. Women sailors—United
States—History. I. Title.
  VB324.W65U735 2015
  359.0082'0973—dc23
                                                                    2015032676

♾ Print editions meet the requirements of ANSI/NISO z39.48–1992
(Permanence of Paper).
Printed in the United States of America.

23 22 21 20 19 18 17 16 15   9 8 7 6 5 4 3 2 1
First printing

CONTENTS

EDITOR'S NOTE

BECAUSE THIS BOOK is an anthology, containing documents from different time periods, the selections included here are subject to varying styles and conventions. Other variables are introduced by the evolving nature of the Naval Institute's publication practices. For those reasons, certain editorial decisions were required in order to avoid introducing confusion or inconsistencies and to expedite the process of assembling these sometimes disparate pieces.

Gender

Most jarring of the differences that readers will encounter are likely those associated with gender. A number of the included selections were written when the armed forces were primarily a male domain and so adhere to purely masculine references. I have chosen to leave the original language intact in these documents for the sake of authenticity and to avoid the complications that can arise when trying to make anachronistic adjustments. So readers are asked to "translate" (converting the ubiquitous "he" to "he or she" and "his" to "her or his" as required) and, while doing so, to celebrate the progress that we have made in these matters in more recent times.

Author "Biographies"

Another problem arises when considering biographical information of the various authors whose works make up this special collection. Some of the selections included in this anthology were originally accompanied by biographical information about their authors. Others were not. Those "biographies" that do exist have been included. They pertain to the time the article was written and may vary in terms of length and depth, some amounting to a single sentence pertaining to the author's current duty station, others consisting of several paragraphs that cover the author's career.

Ranks

I have retained the ranks of the authors *at the time of their publication*. As noted above, some of the authors wrote early in their careers, and the sagacity of their earlier contributions says much about the individuals, about the significance of the Naval Institute's forum, and about the importance of writing to the naval services—something that is sometimes underappreciated.

Other Anomalies

Readers may detect some inconsistencies in editorial style, reflecting staff changes at the Naval Institute, evolving practices in publishing itself, and various other factors not always identifiable. Some of the selections will include citational support, others will not. Authors sometimes coined their own words and occasionally violated traditional style conventions. *Bottom line:* with the exception of the removal of some extraneous materials (such as section numbers from book excerpts) and the conversion to a consistent font and overall design, these articles and excerpts appear as they originally did when first published.

ACKNOWLEDGMENTS

THIS PROJECT would not be possible without the dedication and remarkable industry of Denis Clift, the Naval Institute's vice president for planning and operations and president emeritus of the National Intelligence University. This former naval officer, who served in the administrations of eleven successive U.S. presidents and was once editor in chief of *Proceedings* magazine, bridged the gap between paper and electronics by single-handedly reviewing the massive body of the Naval Institute's intellectual content to find many of the treasures included in this anthology.

A great deal is also owed to Mary Ripley, Janis Jorgensen, Rebecca Smith, Judy Heise, Debbie Smith, Elaine Davy, and Heather Lancaster who devoted many hours and much talent to the digitization project that is at the heart of these anthologies.

Introduction

TODAY WOMEN ARE SERVING in every branch and nearly every component of the armed forces. But this was not always so. For most of their existence, the armed forces were almost entirely the domain of men. Women serving as nurses and in some other limited capacities were aberrations rather than integrated components of the nation's defense. All of that began to change in the latter part of the twentieth century, and it is a process that is nearly complete at the dawn of the twenty-first.

Meeting the challenges of gender integration has been a joint operation that has encompassed all of the armed forces, with the Navy only a part of the larger struggle. But the tribulations encountered by those pioneering women who sought their rightful place in the Navy were unique in many ways, and their story is worth recounting both for its individuality and its importance. From the days of "yeomanettes" to today's Navy—where women command ships and wear admirals' stars—the Naval Institute has chronicled this progression through its books and magazines, creating a historical record and offering the insight that only an open forum can provide.

Selected from a much larger body of work, the articles and excerpts from books in this edition of Naval Institute Chronicles tell a significant portion of an important evolutionary and revolutionary story, one that has had significant impact on the U.S. Navy and will continue to do so in the foreseeable future.

1

"The Eighteenth and Nineteenth Centuries"

Susan H. Godson

(Selection from *Serving Proudly:
A History of Women in the U.S. Navy*,
Naval Institute Press, 2001): 1–31

DURING THE EIGHTEENTH and nineteenth centuries, women sailed in ships of every sort—hospital, whaling, merchant, pirate, privateer, clipper, and war—and often learned the intricacies of shiphandling and seamanship. At the same time, they provided the gentle nursing care so essential to sick and injured men, often using their talents in military settings. These two elements, familiarity with ships and maritime matters and nursing skills, would provide the foundation for women's twentieth-century participation in the U.S. Navy.

1776–1815

Even before there was a U.S. Navy, women had helped gain the country's independence during the American Revolution. Twenty thousand females assisted the Continental Army, some as camp followers, spies, or nurses who provided rudimentary care to the sick and wounded. The legendary Molly Pitcher was a water carrier, and Deborah Sampson enlisted, disguised as a man. Others, in the heat of battle, fired the weapons of their fallen husbands. Several women saw service in the states' navies during the war. In 1776 the Pennsylvania Navy had women onboard some of its galleys; the next year the Maryland Navy's ship *Defence* hired Mary Pricely as a nurse. Even John Paul Jones, when he captured the British sloop *Drake* in 1778, included the cook's wife among his prisoners, taking her onboard his ship, the Continental Navy's sloop of war *Ranger*. Other than these isolated examples, however, women apparently did not play a significant part in naval operations during the American Revolution.

The Continental Navy provided some care for sick and wounded men onboard warships, and the 1775 naval regulations specified that a part of each ship would be set aside as a sick bay. A surgeon, a surgeon's mate, and often crew members would nurse the infirm. Two years later, Congress mandated professional examinations for all naval surgeons and surgeons' mates, a step toward improving the Medical Department.

After the war the Continental Navy disbanded and its ships were sold. The country was left without a navy. By 1794, however, Barbary pirates were prowling the Mediterranean and venturing out into the Atlantic, seizing American merchantmen and their crews. Rather than pay ransom for these prisoners, Congress established a naval officer corps and authorized construction of six frigates to protect American commerce. Each new ship would carry a surgeon onboard. Although the United States and Algiers reached an agreement the next year to end the piracy in exchange for almost $1 million in bribes and naval supplies, Congress begrudgingly authorized completion of three frigates—the *Constitution, United States,* and *Constellation.* Not completed until 1798, these ships became the nucleus of an American naval force hastily assembled during the Quasi War with France (1798–1801). During this crisis, Congress wrenched control of naval affairs from the War Department and established the Department of the Navy in April 1798. The department had two administrative officers—secretary and accountant—and several clerks. President John Adams named Benjamin Stoddert the first secretary of the Navy, and under Stoddert's leadership the Navy grew to fifty-four ships, 750 officers, and 6,000 men.

The end of the Quasi War and the election of Thomas Jefferson to the presidency in 1801 brought a reduction of the American naval establishment to thirteen frigates and one schooner. Jefferson hoped to disband the Navy, but the Barbary pirates dashed this hope by renewing their attacks on American shipping. Then Tripoli declared war on the United States. Determined to defeat the Barbary nemesis, Jefferson dispatched a squadron to the Mediterranean. Following several years of skirmishing, the United States and Tripoli signed a peace treaty in 1805, and American warships soon returned home. Another ten years would pass before the country completely resolved the problem of the Barbary corsairs.

The Quasi War and the need to resist the Barbary depredations of American commerce had given the U.S. Navy permanent status in the eyes of the nation, and with this permanency came traditions and standards for the fledgling naval force. Since 1775 the Navy had had rules that were patterned largely after those of Britain's Royal Navy. In 1802 a new set of regulations detailed the responsibilities of squadron commanders, ships' officers, and crew members. A ship captain's duties included ensuring his vessel did not transport "any woman to sea, without orders from the navy office, *or the commander of the squadron.*" A woman's presence onboard naval vessels must not have been unusual for such a regulation to appear. If so, the American Navy again borrowed customs from the Royal Navy, whose eighteenth-century warships regularly carried wives, "loose women," children, and animals to sea.

As early as April 1802, Commo. Richard V. Morris in the frigate *Chesapeake* applied the regulation when he led his squadron to the Mediterranean to protect American merchantmen. Morris must have granted himself permission, for he brought on the voyage his wife, his infant son, and a black maid. Other wives came, too, and one went into labor. The ship's doctor, hoping to speed the delivery, ordered her placed near the broadside guns and a volley fired. The remedy was successful, and a "son of a gun" was born. Another wife stood in for the mother at the baby's christening, and three other women not invited to the ceremony got drunk in their quarters. Commodore Morris was an ineffective squadron commander (whether because of all the domestic squabbles is speculative); Jefferson relieved him in 1803, and the Navy subsequently dismissed him. Nevertheless, the practice of carrying women in warships continued.

As the problems with the Barbary pirates subsided, other threats to American rights on the high seas increased. Between 1799 and 1812, Britain impressed some ten thousand American sailors and interfered unchallenged with trade. After the unprovoked attack on the *Chesapeake* in June 1807, war fever swept the country, but the United States was woefully unprepared for another confrontation with Britain. President Jefferson had stressed coastal defenses, and the Navy had only two active frigates and about 172 coastal gunboats—hardly sufficient to ward off Britain's massive sea power. Following James Madison's inauguration as president in 1809, economic measures still failed, and the country edged toward another war.

The U.S. Navy would play a crucial part in the coming war, and casualties were inevitable. The Act of 1798 had established a Navy Hospital Fund by deducting twenty cents a month from the pay of merchant seamen, and the next year the benefit expanded to the naval service. In February 1811 Congress passed an act establishing naval hospitals. Secretary of the Navy Paul Hamilton later asked William P. C. Barton, a young naval surgeon, to compose a set of regulations for governing these hospitals. Barton was well aware of the shortcomings in Navy medical care. Shipboard facilities were primitive, and there were no permanent hospitals ashore, only temporary ones in Navy yards. Surely the Navy could provide for its own. As early as 1695 the Royal Navy had emphasized professional care of its sick by building hospitals at Greenwich, Haslar, and Stonehouse; the Continental Army had hospitals with matrons and local women "nurses" during the Revolution.

Borrowing heavily from the work of naval surgeon Edward Cutbush, who had published a book in 1808 on sailors' health and hospital administration, Barton drafted rules for governing naval hospitals. The Navy Department submitted them to Congress in 1812. Each hospital accommodating at least one hundred men should maintain a staff including a surgeon, who must be a college or university graduate; two surgeon's mates; a steward; a matron; a wardmaster; four permanent nurses; and a variety of servants. Barton delineated the duties of each medical care provider. Of particular interest is the role of matron, who ideally, Barton noted, should be the steward's wife. She would be responsible for keeping patients and wards clean, carrying out the surgeon's directives, and supervising nurses and servants. Barton's report did not specify whether nurses were to be male or female and made little mention of duties, except that one should avoid displeasing the hospital surgeon. An addendum recommended specific rank for surgeons and the same pay and emoluments as Army surgeons received. As for building sites, eight to ten acres were necessary for each hospital, and at least fifteen acres if an asylum was added.

Not satisfied with the hastily drafted suggestions, Barton expanded his theories in a treatise published in 1814. In discussing his proposed internal organization and government of marine hospitals and staff duties, Barton further defined the matron's characteristics: she should be "discreet . . . reputable . . . capable . . . neat, cleanly, and tidy in her dress, and urbane and tender in her deportment." She would supervise the nurses and other attendants as well as those working in the laundry, larder, and kitchen, but her main function was to ensure that patients were clean, well-fed, and comfortable. Barton made no mention of the matron's providing any type of medical care.

Barton's treatise increased the responsibilities for nurses, now described as "women of humane dispositions and tender manners, active and healthy . . . neat and cleanly . . . and without any vices of any description." They would dispense medicines and diets to the sick, keep the wards clean, and watch the sick—even sitting up with them at night. In addition, there should be "orderly-men," or male nurses. These men were to assist the female nurses and could perform duties in caring for the sick "that women could not decently attend to."

Although Barton showed foresight in proposing the extensive use of women in caring for the sick, he envisioned their filling the customary nurturing roles in society: keeping the infirm clean and comfortable and following doctors' orders. He made no mention of any specialized training or education for these "nurses," which is not surprising. In 1814 there was no professional training for nurses in the modern sense. Barton's nurses would merely perform women's usual duties in a military setting. Unfortunately, Barton's suggestions for female nurses did not materialize.

Two years before Barton's *Treatise Containing a Plan for the Internal Organization and Government of Marine Hospitals* was circulated, the United States and Great Britain were at war. After a number of single-ship victories, the U.S. Navy found itself boxed in by a Royal Navy blockade of the East Coast. Commo. Stephen Decatur planned to run his squadron of frigates *United States* and *Macedonian* and sloop-of-war *Hornet* from New York through the blockade in May 1813. Believing there would be heavy casualties, Decatur took two women onboard the *United States* as "nurses." Both were wives of crewmen and had no medical training. Mary Allen accepted Decatur's offer of accompanying her husband, John, on condition that she serve as a nurse. The second woman, Mary Marshall, the wife of another seaman, also acted as a nurse. The ship's log for 10 May 1813 lists the women as supernumeraries.

Although Decatur never broke through the blockade, he kept the "nurses" onboard for months. Mary Allen's naval service ended in October, when her husband drowned and she secured permission to leave the ship. Mary Marshall might have remained on *United States* until her husband was reassigned the following spring, or she might have left with Mary Allen. These two women, the first documented hired "nurses" serving on a U.S. Navy warship, were early predecessors of nurses in ships during the Civil War and the Spanish-American War.

The War of 1812 inspired another form of feminine patriotism: a woman disguising herself as a man in order to serve on a U.S. Navy warship. Ever since the alleged voyages of Lucy Brewer in the *Constitution*, naval historians have had to confront the persistent story of her service and have been thwarted by a lack of hard evidence either supporting or refuting the tale.

True or not, Brewer's exciting adventures had ample precedent in European, especially British, navies and in popular literature and song. Appealing primarily to lower-class women, British ballads extolling female warriors enjoyed great popularity from the mid-seventeenth century to Victorian times. Typically, the "warrior" had been separated from her man by war or by her irate father, dressed as a male, and went to sea or to war. She was always an exemplary, fair, and virtuous heroine.

There was just enough fact to sustain these stories and songs. In the eighteenth century, the Royal Navy carried a marine named William Prothero in the *Amazon*. William turned out to be an eighteen-year-old Welsh girl. Another female Royal Marine sustained twelve wounds while fighting against the French in India. Among the most famous of these transvestite warriors was Hannah Snell, who served in both the British army and navy from 1745 to 1750 as James Gray. After she retired with a government pension, Snell's exploits became widely known through her biography, *The Female Soldier*. Equally renowned was Mary Ann Talbot, who published her own memoir of her service as a servant, a drummer, and a sailor in a French privateer and finally in three British warships. She, too, collected a pension at retirement. Even more relevant for American women were the Revolutionary War exploits of Deborah Sampson, which were published in 1797 as *The Female Review*. The Massachusetts native would be a compelling role model for Lucy Brewer.

According to her autobiography, Lucy left her family's home forty miles west of Boston when she was sixteen years old, after being seduced and betrayed by a lover. Not wanting to disgrace her family, the pregnant girl quickly found work in a Boston brothel. Her baby died at birth, and Lucy continued in the world's oldest profession for three more years. In 1812 a young naval lieutenant, presumably a customer, told Lucy the story of Deborah Sampson, suggesting that Lucy disguise herself as a man and sign on to the frigate *Constitution*, then in port. Living in a busy seaport, Lucy undoubtedly would also have been familiar with barroom ballads about female warriors.

Lucy dressed in a sailor's suit and volunteered as a marine in *Constitution*. In those days many young boys served in various capacities onboard American ships, and the average sailor was relatively small. Lucy would have had little trouble passing as a male. She learned to handle firearms as well as any recruit and took part in three major engagements. After the War of 1812, Lucy reverted to feminine dress, returned to her parents' home, and wrote her memoirs, which went to three printings.

Lucy's memoir is the only written account of her adventures. There is nothing about them in any official documents. Ignored or dismissed by most naval historians, the story has lingered with astonishing persistency and was the topic of a book published during the country's bicentennial. Fictional or not, the story of Lucy Brewer left an enduring legacy of a woman serving in a combatant ship.

1815–1860

As Lucy Brewer entered married life, she became a part of the cult of domesticity, a trend that lasted throughout the nineteenth century. In sharp contrast to earlier times, when they had filled skilled occupations as school mistresses, shopkeepers, and silversmiths, women were increasingly denied jobs, training, and education. Urbanization and industrialization caused men to move from rural and small-town America to the cities and to more professional occupations, a trend that left many middle- and upper-class women isolated within their homes and bound by their duties as wives and mothers. Markedly influenced by European ideas of womanhood, "fashionable" American women accepted their place in the domestic sphere. Cautioned against any pursuits outside the home that might cause the loss of "femininity," women's only public role was through their husbands and families. After all, women's "special" nature endowed them with patience, endurance, passive courage, and higher moral qualities, such as piety, submissiveness, and nurturance, attributes that made women supremely fit to reign within the domestic sphere.

Social reform movements, which began in the 1820s, allowed proper women to contribute their special talents to the outside world. Led by churches and benevolent societies, movements and their causes proliferated. Here, certainly, were respectable outlets for women's energies. Within reform groups women demonstrated their leadership capabilities and formed associations that trained them for larger roles in society.

One such reform movement was evangelical Christianity, which preached the urgent need for salvation, as manifested by good works, temperance, and the end of corporal punishment. The U.S. Navy was a ready target. To reform and uplift sailors, the Boston Society for the Religious and Moral Improvement of Seamen was established in 1812, followed five years later by the Marine Bible Society. These two societies soon spawned such auxiliaries as the Female Seamen's Friends societies. Founded in 1820, the Society for the Relief of Families of Sailors Killed While in Service was the forerunner of the Naval Relief Society of 1904. In 1826 the American Seamen's Friend Society organized with branches in port cities in the United States and abroad. It spawned the Seamen's Bank for Savings and published the widely read *Sailor's Magazine*. These groups tried to improve sailors' living conditions ashore by setting up boardinghouses (alternatives to brothels) and encouraging savings. They also maintained employment registries, libraries and reading rooms, and training schools and sent chaplains to foreign ports. The reformers were instrumental in the drive to abolish flogging in the Navy and in commercial vessels. This inhumane punishment ended in 1850. Drunkenness, the major cause of flogging, decreased as the Navy slowly reduced its grog ration, but it was 1862 before Congress prohibited alcoholic spirits on board naval ships.

Although women were active in the American Seamen's Friend Society, they were frozen out of most leadership roles in major reform movements such as abolition (it was not ladylike to speak in public), and their anger and resentment mounted. Finally, in 1848, the first Women's Rights Convention met at Seneca Falls, New York. Its 240 delegates, led by abolitionists Elizabeth Cady Stanton and Lucretia Mott, enumerated their grievances, which had resulted from men's "absolute tyranny" over women. In the Declaration of Sentiments and Resolutions, they demanded equal moral, political, and economic rights and privileges. Most people regarded these early feminists as radicals, and they attracted only a scant following. They did, however, challenge the prevailing ideology that women's only place was in the home, and they paved the way for the nascent antebellum women's rights movement of the 1850s. But most women remained contentedly in their defined sphere; it would take a war to release them from the cult of True Womanhood.

While women debated their rights, or lack thereof, the United States enjoyed more than thirty years of peace. The Navy's primary missions became protecting American commerce against pirates, showing the flag around the globe, opening new markets, and charting seas and coastlines. To achieve these goals, Congress authorized increased ship construction; and the Navy established the Pacific, Brazil, West Indies (later, Home), East Indian, and African Squadrons in addition to its old Mediterranean Squadron. Worldwide expansion and far-flung stations required better administrative procedures, first overseen by the Board of Navy Commissioners (1815) and then by five naval bureaus (1842). By the early 1840s, a new generation of naval leadership grudgingly accepted such innovations as steam-powered warships, and the Corps of Engineers was formed.

During these same years, naval leaders instigated means for improving officers' professional standards. To disseminate knowledge, they established the U.S. Naval Lyceum (1833), the *Naval Magazine* (1836), and the American Historical Society of Military and Naval Events (1836). The Navy Department set up the Depot of Charts and Instruments, forerunner of the Naval Observatory and Hydrographic Office. In 1845, Secretary of the Navy George Bancroft convinced Congress to establish the U.S. Naval Academy for training potential officers. Ten years later, the Naval Efficiency Act introduced a retirement process for incapable officers. In 1861 all had to retire at age sixty-two or with forty-five years' service.

In military operations, the Navy helped remove the Seminole Indians from the Florida Everglades in the mid-1830s, getting its first taste of riverine warfare. When the Mexican War began in 1846, the Navy blockaded ports in the Gulf of Mexico, staged the amphibious invasion of Veracruz, and played a major role in the conquest of California. As peace returned, the Navy continued opening foreign markets for American commerce. In the mid-1850s Commo. Matthew C. Perry, through astute diplomacy and a grand display of naval force, opened the door for trade with Japan. Simultaneously, Secretary of the Navy James C. Dobbin convinced Congress to launch a shipbuilding program, and by 1860 there were thirty-eight steam vessels and forty-four sailing ships in the fleet.

As the Navy itself underwent change and growth, so did its Medical Department. After passage of the act establishing naval hospitals in 1811 and William P. C. Barton's ambitious plans for them, little progress occurred for twenty years. The Navy maintained temporary, and primitive, hospital facilities in Navy yards at Washington, Philadelphia, New York, and Boston. During the 1820s the Navy acquired the necessary land, and in the following decade built permanent hospitals at Portsmouth, Virginia; League Island, Philadelphia; Portsmouth, New Hampshire; Chelsea, Massachusetts; Pensacola, Florida; and Brooklyn, New

York. The hospital at Philadelphia also had an asylum for disabled veterans. Later hospitals followed at Newport, Rhode Island; Washington; and Annapolis. No one suggested matrons or women nurses for these hospitals.

Afloat, surgeons served in major warships around the globe. Although separate hospital ships had been recommended as early as 1808, none was built. The ship's sick bay continued to be the place of patients' recovery. The surgeon, his assistant, and crew members called loblolly boys (possessing no medical training or experience) provided rudimentary care.

In an early, short-lived attempt at improving the professional quality of medical personnel, physician Thomas Harris set up a school in 1822 to teach new medical officers about naval hygiene, customs and usage, and the military. Another stride toward better medicine was the Navy Laboratory, opened at the Naval Hospital at Brooklyn in 1853. The lab provided pure drugs, especially the new ether and chloroform, to the Medical Department. In the 1870s the Naval Hospital at Brooklyn briefly gave a two-year course in naval medicine, and in 1893 a similar three-month program began at the Naval Laboratory and Department of Instruction at Brooklyn. These were the forerunners of the Naval Medical School, established in Washington, D.C., in 1902. Other means of upgrading the professional quality of naval medical personnel were examinations for surgeon's mates in 1824, for assistant surgeons in 1828, and for surgeons in 1829.

An advance for naval medicine came with the Navy's establishment of the Bureau of Medicine and Surgery in 1842. Named chief of the bureau was surgeon William Barton, who, thirty years after devising a plan for naval hospitals, had the opportunity to direct the Navy's medical program. During his two years as chief, Barton insisted on high professional standards for the physicians of the Medical Corps, advocated better recruiting procedures, established medical libraries at each naval medical facility, supplied naval surgeons with well-known medical journals, and advocated abolishing liquor onboard ships.

Barton's grand plans were limited by congressional action in 1842, which capped the number of surgeons at sixty-nine and the number of assistant or passed surgeons at eighty. In 1846 Secretary Bancroft awarded military rank to medical officers. Although the Navy grew during the next fifteen years, the number of medical personnel remained fixed, and by 1860 fully one-sixth were too old to serve afloat.

Like medicine itself in the antebellum era, naval medical practices showed sporadic and uneven advances. It was an age when diseases such as scurvy, smallpox, cholera, and yellow fever ran rampant and standard treatments consisted of bloodletting, purgatives, and emetics.

Until 1847 there were no female health-care professionals, that is, women trained as nurses or doctors. That year one woman, Elizabeth Blackwell, was admitted to the all-male Geneva Medical College in western New York and became America's first formally trained woman doctor. Medical schools remained male bastions, however, so several such colleges for women opened in the 1850s. But with the exception of Roman Catholic religious orders, especially the Sisters of Charity, which taught their own members how to care for the sick and the poor, most people thought there was no need for either female doctors or nurses because women without special training had always taken care of the sick. Some basic instruction began in 1839 at the Nurse Society of Philadelphia, which taught lay women how to help with home maternity cases. Little else was done.

Suddenly, the Crimean War shown a spotlight on the need for nurses, and one woman's efforts in that war would revolutionize nursing. In 1854, Great Britain, France, the Ottoman Empire, and Sardinia allied to halt Russian expansion in the Black Sea area, and the Crimean Peninsula became the battleground. As British army casualties mounted, many wounded and dying soldiers suffered from lack of medical care and supplies, so the British secretary of war asked Florence Nightingale, daughter of a wealthy and distinguished family, to become general superintendent of nursing for the Army hospitals in Turkey. Nightingale had studied nursing in Kaiserwerth, Germany, and in France (unheard of for a young woman of her social standing) and had become superintendent of a women's hospital in London. After recruiting thirty-eight other nurses, she went to Scutari, where the deplorable, filthy Turkish barracks serving as a hospital for British troops needed her energetic direction. Responsible for the nursing system, Nightingale also supervised the diet kitchens, laundries, and sanitary facilities and obtained supplies and laboratory equipment. Sent in early 1855 to report on and rectify the disease-spawning conditions at Scutari, the Sanitary Commission soon joined in the cleanup.

Although administrative duties took up most of her day, Nightingale trudged through miles of hospital corridors at night to nurse the ill and wounded soldiers, who fondly named her "the Lady with the Lamp." The British soon put her in charge of nursing at all military hospitals in Crimea, and the death rate there fell from 42 to 2.2 percent. Thanks to Nightingale's heroic efforts, modern military nursing had begun, in turn leading to the development of professional nursing. After the war a grateful public gave her funds, and in 1860 she established the Nightingale Training School for Nurses at London's St. Thomas's Hospital. Nurses would be taught and then sent to outlying hospitals to teach others. The control and instruction of nursing was to be handled by women nurses, not by male doctors, a novel concept for the times. For generations, the "Nightingale System" would influence nursing education in Europe and the United States.

Before the Crimean War, neither nurses nor any other women had a place in the U.S. Navy, but there were activities involving women with the Navy. In 1827 the first recorded instance of a woman's christening a ship occurred when "a young lady of Portsmouth" christened sloop-of-war *Concord*. No one knows her name. In 1846 Lavinia Fanning Watson of Philadelphia christened sloop-of-war *Germantown* at the Philadelphia Navy Yard. Women sponsored more and more ships as the nineteenth century went on, but men also continued to christen vessels.

Women, and often children, had always taken passage in naval ships when there was sufficient cause for transporting them. The most common reason was carrying diplomats and their families to foreign posts, a practice that continued throughout the nineteenth century. For transporting passengers, naval regulations had always required authorization from the Navy Department, secretary of the Navy, or the fleet or squadron commander in chief when in foreign waters. Such passengers were not to interfere with operation of the ship.

Officers' wives presented another set of problems. Ever since Commodore Morris's chaotic 1802 voyage in *Chesapeake,* with a large complement of wives onboard, the Navy had slowly tightened the rules about carrying families. Sailors complained about the women, and a young midshipman in *United States* in 1832–33 wrote in his diary that the officers had wished Commo. Daniel Todd Patterson's wife and daughter home a thousand times. The final straw came when Commo. Isaac Hull, commanding the Mediterranean Squadron in 1838, took his wife Ann and her sister Jeannette Hart to sea onboard *Ohio.* Ann had always accompanied Hull during their twenty-five years of marriage, but Jeannette proved a disruptive influence. The wardroom officers rebelled against the women's presence, and riots, desertion, and insubordination marred the cruise.

In 1842 Congress passed a joint resolution asking Secretary of the Navy Abel P. Upshur to prepare new rules and regulations for the U.S. Navy. Among Upshur's recommendations was a specific prohibition: "In no case shall a commander-in-chief, or commander of a vessel, take his wife to sea in a public ship." Finally, by 1881, all women were barred from either residing on or taking passage in any naval ship.

Women may have worn out their welcome in naval ships, but they continued a long tradition of sailing on pirate, privateer, clipper, and whaling vessels. The most flamboyant were the lady pirates, who left a trail of robbery and murder rivaling that of their male counterparts. In American piracy, for example, Rachel Wall and her husband, George, operated off the New England coast in the late eighteenth century. Their modus operandi called for taking their vessel into the shipping lane after a storm while feigning damage and distress. Well-meaning traders stopping to offer assistance promptly had their throats slashed and their money and merchandise stolen. Afterward, the Walls sunk the rescue ships. Their profitable business continued until George misjudged a storm's fury, was swept overboard, and drowned. Rachel settled in Boston, where she worked as a servant while moonlighting in petty larceny. She was hanged in 1789 for robbery.

The same techniques, and often the same crews, used in these high-seas robberies appeared in privateering, the government-sanctioned version of piracy. Just before the American Revolution, Fanny Campbell of Lynn, Massachusetts, disguised herself as a man and became second officer in the British brig *Constance*. Fanny led a mutiny and became commander of the brig. Having stolen the ship, she and the crew were now pirates. After *Constance* encountered and captured several British ships, the female commander returned to Marblehead with her prizes. Along with her new husband, William Lovell, and most of her crews, she received a commission as a privateer. Fanny immediately retired to raise her family and left the privateering to William for the rest of the war.

Women onboard merchant ships were not as unusual as female pirates and privateers, and during the nineteenth century it became standard practice for captains to take their wives and families to sea. Many wives learned navigation and were able to chart the ship's course; others were proficient in bookkeeping and served as pursers. Captains' children who had grown up at sea learned these same skills; the sons also mastered shiphandling, and some daughters later became captains themselves.

Most glamorous of the nineteenth-century merchant vessels were the giant clipper ships that raced around Cape Horn to California or China. During the 1850s women briefly commanded two of these. Hannah Burgess of Boston, wife of William, who was captain of the large clipper *Challenger*, went to sea with him in 1855. She was an expert navigator. The following year, when William fell ill and died, she commanded the ship for three weeks until she made port at Valparaiso. That same year another New England wife, Mary Patten, accompanied her husband, Joshua, captain of the clipper *Neptune's Car*. After they sailed from New York, Joshua fell critically ill with brain fever. The first mate was in the brig, so Mary, pregnant and only nineteen years old, took command of *Neptune's Car* and after fifty-two harrowing days, got her around Cape Horn to San Francisco. The ship's owners rewarded Mary's heroism, giving her one thousand dollars for bringing their ship and its cargo to safety.

During these years, whaling vessels prowled the seas searching for the mammals that provided oil for machinery and lamps and whalebone for clothing and household use. These voyages often lasted four or five years, such separations playing havoc with family life. Whaling captains began taking their wives and families to sea, and the ships with women onboard became known as "petticoat whalers." The trend grew rapidly, and by midcentury about 20 percent of whaling captains were bringing their families with them. These families often kept diaries and journals, which have provided abundant information about petticoat whalers. The most famous whaling wife was Mary Brewster, who sailed with her husband, William, in Tiger in 1845. She went around Cape Horn and cruised off the California coast, reaching such exotic places as Samoa, the Cook Islands, and the Sandwich (Hawaiian) Islands. In 1849 she was the first American woman in the western Arctic.

On shore, however, the whaling industry was changing. When Nantucket had been the center of whaling, wives left behind demonstrated an independence enabling them to handle business matters and provide for their families. By the 1830s the center had moved to New Bedford, where the change brought a melding of maritime and land affairs into the pattern of industrialization. In this milieu, women had less opportunity to act independently.

The Civil War, 1861–1865

When the Civil War tragically divided the nation in April 1861, the U.S. Navy had two primary missions: blockading the Southern coasts and controlling the Mississippi River and its tributaries. In carrying out these tasks, Secretary of the Navy Gideon Welles oversaw the expansion of the Union Navy from 90 vessels, 1,300 officers, and 7,600 men in 1861 to 671 vessels, 6,700 officers, and 51,500 men by late 1864. To help administer the growing Navy, the number of naval bureaus increased to eight in 1862.

As war continued, casualties mounted on both sides. By 1865, 618,000—360,000 Union and 258,000 Confederate troops—had died from battle or disease. Evacuation of the sick and injured by ship, as well as by railroad and wagon, became commonplace. Both on land and on water, women played an increasingly visible and effective role in caring for these casualties.

Anticipating a great need for medical care, Secretary of War Simon Cameron appointed Dorothea L. Dix superintendent of the Army nurses. Dix, a Boston schoolmistress, had worked diligently during the 1840s and 1850s for reforming the treatment of the mentally ill. Dix set strict standards for Army nurses: they must be between thirty-five and fifty years of age, plain looking, and possessed of good morals. Only brown, gray, or black clothing without ornamentation was acceptable. Dix sent about one hundred women volunteers for one month's orientation at Bellevue Hospital in New York; it was their only medical training.

Eventually, Dix appointed 3,214 of the approximately 9,000 nurses in more than two hundred of the Army's permanent and field hospitals. These untrained nurses performed invaluable service caring for the sick and wounded under primitive and grueling conditions. Women volunteered for nursing duty in the South as well, although cultural beliefs about such activity limited their numbers to about 1,000 who served in hospitals. Countless others nursed Confederate casualties in private homes.

Still another woman, Clara Barton, directed nursing as well as independent relief operations, first from Washington, D.C., later in South Carolina. She collected and carried wagon loads of supplies directly to battlefields, nursed and comforted the wounded and dying, fed the hungry, and assisted in medical care and procedures. The philanthropic work of this "Angel of the Battlefield," as Barton became known, would have lasting consequences.

Doctors and military men were bitterly opposed to the presence of these nurses because they did not believe ladies should be tending male patients. Nurses often discovered and exposed gross malfeasance, dishonesty, and incompetence among hospital staff members, who, in turn, determinedly tried to rid the Army of these nuisance women. Military surgeons sometimes met their match when pitted against such a formidable woman as Mary Ann "Mother" Bickerdyke, who traveled from one battle zone to another organizing hospitals, enforcing standards of cleanliness, ministering to patients, and supervising diet kitchens.

Early in the war it became apparent that the Army Medical Department would be inadequate to handle the growing carnage on the battlefields. In helping alleviate the inevitable suffering and death, Northern women expeditiously set up soldiers' aid societies and the Women's Central Association of Relief in New York City. The U.S. Sanitary Commission, established in June 1861, coordinated these relief efforts. Southern women also organized aid societies, but on a local rather than regional basis. Borrowing from British experience in the Crimean War and their subsequent sanitary commission sent to improve army hospitals in Crimea, the U.S. Sanitary Commission initially dispatched physicians to investigate conditions in U.S. Army hospitals and camps. The commission's scope widened, and it became the largest welfare organization of the war. Although men were involved in the commission, most members and leaders were women. The group distributed food, clothing, and medical supplies, raised funds, and served as hospital nurses or aides.

As the fighting moved to the Virginia Peninsula in 1862, the Sanitary Commission converted some idle transport steamboats into floating hospitals for evacuating the wounded. The commission paid the cost of refitting the ships and staffed them with surgeons, dressers (often second-year medical students), male and female nurses, and various servants. Women onboard prepared food, stocked shelves, and made the infirm as comfortable as possible. After evacuating about eight thousand men from the peninsula, the commission's hospital transports carried the sick and wounded along the Atlantic coast throughout the war. The commission also sent supplies and delicacies for the sick to at least twenty-three naval ships and furnished medical supplies for a fleet hospital at Pilot Town near New Orleans. Although the commission had offered its services to the Navy, that arm of the military had no need of their transports in the East. The Navy's ordnance ship *Ben Morgan* served as an emergency hospital off Hampton Roads, Virginia, from February to June 1862. There were no other hospital facilities afloat for the North Atlantic Squadron except ships' sick bays.

In the West a separate organization performed relief services. Established in September 1861 and based in St. Louis, the Western Sanitary Commission assisted armies near the Mississippi River. It set up four large hospitals, started soldiers' homes, fitted out hospital transports, and sent medical supplies to naval gunboats. As in the East, the Western Sanitary Commission chartered steamers for transporting the wounded to shore hospitals. Women served as nurses onboard these ships, and the indomitable Mother Bickerdyke took charge of the *City of Memphis*.

The most effective female nurses in the commission and in the military were the more than six hundred sisters from twelve Roman Catholic religious orders, especially the Sisters of Charity and the Sisters of the Holy Cross. Although they had no formal nursing training, they were accustomed to strict discipline and obedience, so the sisters readily fit into the male-dominated medical and military worlds. On 21 October 1861

Indiana governor Oliver P. Morton appealed to Father Edward Sorin, superior of the Sisters of the Holy Cross, for nuns to serve as military nurses. The next day Mother M. Angela and five others left for the hospital at Cairo, Illinois, and then went to the Army hospital at Mound City, Illinois. Eventually, about eighty members of this order served as nurses in military hospitals.

On 7 April 1862 a federal gunboat captured the 786-ton Confederate sidewheel river steamer *Red Rover* at Island Number Ten on the Mississippi. The Army had her refitted as a hospital ship, and the Western Sanitary Commission furnished thirty-five hundred dollars in medical supplies and provisions. The new floating hospital was the most modern of its kind: it had a 300-ton-capacity ice box, bathrooms, a laundry, an elevator, nine water closets, gauze window blinds, two kitchens, and a "regular [male] corps of nurses." George H. Bixby was the senior surgeon onboard. The *Red Rover* went into service in June, and among her first patients were men wounded in an explosion on the gunboat *Mound City*. Suffering from severe burns and scalding, the injured were taken to the hospital at Mound City. Seeing victims of the *Mound City* disaster, Sister Athanasius, a nurse at the Mound City hospital, volunteered for duty in *Red Rover* and spent the rest of the summer there. Mother Angela offered to provide nurses for the Navy, an offer the Navy later accepted.

In September *Red Rover* arrived at St. Louis for fitting out for the winter, and Sister Athanasius left to take charge of a Washington hospital. The *Red Rover* had been a captured vessel, so the Illinois prize court sold her to the U.S. Navy on 30 September for $9,314. The Navy had its first authentic hospital ship. The next day the western flotilla gunboats under Army control were put under naval command and became the Mississippi Squadron, which now included *Red Rover*.

The *Red Rover* remained at St. Louis for three months, and while being refitted, she took on patients from the overcrowded Mound City hospital. Ordnance steamer *Judge Torrence* also served as a temporary hospital until *Red Rover* was ready. Finally, *Red Rover* was commissioned in the U.S. Navy on 26 December 1862. She carried twelve officers, thirty-five enlisted men, and about thirty people in the medical department.

Among the medical personnel were four nuns from the Sisters of the Holy Cross, the same order that housed Sister Athanasius, the first nurse in *Red Rover*. On Christmas Eve, Sisters Veronica, Adela, and Callista came onboard, and in early February, Sister St. John of the Cross joined them. Five black women—Alice Kennedy, Sarah Kinno, Ellen Campbell, Betsy Young, and Dennis Downs—assisted the nuns. During the course of the war, more than twelve members of this religious order served in *Red Rover*, and Sisters Adela and Veronica remained onboard the entire time.

The *Red Rover* sailed on 29 December 1862 and became a familiar part of the Mississippi Squadron. Always near the flagship and close to the scene of operations, *Red Rover* carried 2,497 patients, both Union and Confederate, to shore hospitals. The Sisters of the Holy Cross alleviated the suffering of the sick and wounded as well as they could. They even prepared delicacies such as herb tea, meat broth, and custard for men too ill for standard rations of cold beans and hardtack. The nuns demonstrated both women's ability to serve during wartime in a naval ship and the advantages of the then fashionable womanly virtues of gentleness and compassion. Unintentional pioneers, these Catholic sisters were, in effect, the forerunners of the Navy Nurse Corps. In gratitude Congress belatedly voted a pension for them in 1892, but only sixty-three were still alive. In 1906 the Grand Army of the Republic awarded the bronze medal to the nineteen surviving sister-nurses.

As war slowed on the Mississippi after the fall of Vicksburg in July 1863, *Red Rover* continued her river trips. She tied up permanently at Mound City in late 1864. Surgeon Bixby and two sisters remained onboard until the ship was decommissioned on 17 November 1865. She was soon sold at public auction for forty-five hundred dollars. The Navy's use of a fully equipped floating hospital at the scene of operations had been a success. Except for sailing ship *Idaho*'s brief stint as a hospital and store ship for the Asiatic Squadron in 1868–69, there would be no other hospital ship until the Spanish-American War.

During the Civil War, nurses and all other medical personnel had more work than they could handle, for casualties included more victims of disease than of gunfire or explosions. The leading causes of death were dysentery, typhoid, malaria, pneumonia, and smallpox. Sweltering heat in the Mississippi area exacerbated any medical condition, and the still primitive state of medical care guaranteed a high mortality rate.

To relieve overcrowding at Army hospitals, the Navy established a temporary hospital in a Memphis hotel and later opened another hospital at New Orleans. But naval hospitals provided rudimentary care at best. One young acting ensign, Robley D. Evans, wounded at Fort Fisher, described the sordid conditions at the Norfolk Naval Hospital. Riddled with vermin, the hospital had no modern equipment or trained nurses. A diet of bacon and cabbage was meant to sustain the patients. Evans fought off the surgeon's attempt to amputate his shattered legs, developed an infection and a fever, and nearly died. After months in the hospital, he recovered sufficiently (in spite of the medical attention) to go home to Philadelphia.

Nurses had played an important role during the Civil War, but other women found different ways to assist the U.S. Navy. On 10 January 1863 the Confederate privateer *Retribution* captured the USS *J. P. Ellicott*, a brig sailing near St. Thomas. The *Retribution* took onboard *Ellicott*'s officers and crew and later deposited them on the island of Dominica. Overlooked was the mate's wife; she was left on *Ellicott* to serve refreshments to *Retribution*'s prize crew. Allegedly, her "refreshments" got the prize crew so drunk that she was able to tie them up below deck. Assuming sole command of *Ellicott,* she navigated the brig into St. Thomas and turned her prisoners and the ship over to naval authorities.

Some women helped Union naval operations as lighthouse keepers. Harriet Colfax, a wartime nurse, also served as keeper of the Michigan

City, Indiana, light for many years. Martha Coston, widow of naval scientist Benjamin Franklin Coston, finished developing her husband's flare signaling device and a signal code. She sold her patent to the government when the war started, and the "Coston light" enabled blockading Union ships to signal each other during the night and at long distances.

As in earlier wars, many women disguised themselves as men, serving in both the Union and Confederate armies, and some later received pensions. Others helped military operations as spies, scouts, saboteurs, messengers, couriers, and blockade runners. Many were cooks, laundresses, and sutlers. Still others worked in munitions plants. A few, such as Mary E. Walker, were licensed doctors. Walker was an assistant surgeon in the Army of the Cumberland and later received the Medal of Honor.

On the home front, women replaced men in factories and in teaching. For the first time, hundreds took government jobs as clerks in the Treasury, Post Office, and War Departments as well as with the Confederate government. In 1863 bread riots in the South and antidraft riots in the North brought thousands of women into the streets. Leaders of the earlier women's rights movement banded into the Women's National Loyal League and gathered four hundred thousand names petitioning Congress to abolish slavery.

All these different means of aiding the war effort, in addition to women's work in the U.S. and Western Sanitary Commissions and in nursing, opened new doors. Women had actively participated in the war, gained leadership experience, and honed their organizational skills.

1865–1898

Except for a war scare in 1873, when the Spanish seized the *Virginius*, the international scene was relatively calm. By 1880 the fleet had dropped from about 626 commissioned vessels to 48 outdated ships. Once again sail became important, for two reasons: steam plants in ships continued to work inefficiently and the United States did not have enough coaling stations abroad to sustain the steam engine. The officer corps dwindled and promotions were slow; enlisted men numbered only six thousand, and most were foreigners. Simultaneously, the Medical Department reflected the same stagnation and decline. It did, however, open a naval hospital at Yokohama, Japan, in 1872—the only such American facility in the Far East. Nevertheless, the Navy kept showing the flag around the world, and the United States and Korea signed a trade treaty in 1881. Exploration of uncharted seas continued.

The country would need the Navy as industry and agriculture then undergoing an explosive growth looked abroad for markets. Such foreign trade required a fleet for protection. President James A. Garfield's secretary of the Navy, William H. Hunt, began campaigning in 1881 for a revitalized Navy; and his successor, William E. Chandler, pursued the modernization goal. Two years later Congress appropriated $1.3 million for the first four ships of a new steel Navy. Two more cruisers came next, followed by the battleships *Texas* and *Maine*.

Fueling the drive for expansion and modernization was the impact of Alfred Thayer Mahan and his book, *The Influence of Sea Power upon History, 1660–1783* (1890). Mahan demonstrated that control of the sea had created mighty empires in the past and called upon the United States to build a powerful battleship fleet and acquire overseas bases. Mahan's theories gained worldwide praise and meshed with the rising tide of imperialism at home. President Benjamin Harrison's secretary of the Navy, Benjamin F. Tracy, adopted the trend, and, at his urging, Congress authorized three more battleships in 1890. The nascent attempt to bring the American Navy on a par with European nations succeeded, and by 1897 the Navy had risen to respectability among the world's fleets. It was an effective fighting force of fifty-four ships, sixty-four auxiliary vessels, and twenty subsidized steamers. Five battleships, sixteen torpedo boats, and one submarine boat were under construction. At the same time, the Navy continued its drive for professionalization. New offices—the

Navy Department Library, Naval Intelligence, and Naval War Records—opened in the 1880s. In 1884 the Naval War College introduced its postgraduate program for officers.

The Medical Department, as well, had kept pace with naval expansion, and throughout the century it had strived to professionalize its services. Although it had difficulty attracting and retaining medical officers, it maintained high standards for those entering the service. In 1893 the Naval Laboratory and Department of Instruction established a course required for junior medical officers before they began active duty.

During the 1890s the Medical Department repaired, rehabilitated, and modernized naval hospitals, making them comparable to good civilian hospitals. By 1897 all naval hospitals had new aseptic operating rooms with modern furniture and appliances, bacteriological and chemical laboratories, disinfecting plants, electricity, X-ray machines, and adequate sanitary facilities. Each had an ambulance available.

Aware that the practice of using seamen as nurses was unsatisfactory, the surgeon general had consistently called for a hospital corps, composed of hospital apprentices, apprentices first and second classes, and pharmacists. The U.S. Army and foreign countries and their navies had had such corps for years. Training schools for apprentices would operate at naval hospitals, so there would be an ample supply of trained "nurses" for both hospital and ship duty. The surgeon general recommended that ambulance (hospital) ships be available for all fleet actions.

There was no place in this new Navy for women, but they made inroads as civilians connected with the Navy. By 1869 they worked as laundresses, cooks, and cleaning women at the Naval Asylum in Philadelphia. Before the end of the century, there were women working at the Navy's clothing factory in Brooklyn, New York. Others provided such piecework service as quilting for naval needs at the Washington Navy Yard.

Although the Civil War had provided women access to jobs in government departments, the Navy was slow in hiring female clerks. One reason was that some men believed these working women had acquired the reputation of "brazen hussy" during the war. By 1879, at least one woman, M. M. (Martha M.) Smith of New York, was working as a "writer" in the Hydrographic Office for seventy-five dollars a month. The Navy hired more female writers in the same office; copyists, clerks, then typists in the judge advocate general's office; clerks in seven naval bureaus; and a telegraph operator in the secretary of the Navy's office. The Hydrographic Office employed several women draftsmen, a stenographer, a custodian of archives, an engraver, and even a laborer. After the Navy established the Library and Naval War Records Office in 1882, women readily found jobs there as clerks. A German-born woman became a translator in the Office of Naval Intelligence, and the Navy Pay Office in Philadelphia hired a woman stenographer. Later, the Marine Corps employed a woman clerk. By the end of the nineteenth century, at least thirty-six women worked in a clerical capacity for the naval shore establishment, primarily in Washington.

Navy Department civilian workers represented only a fraction of the women escaping the confines of their domestic life during the last third of the nineteenth century. The women's rights movement, quiescent during the war, split into two organizations in 1869 because of differences in suffrage goals. To pursue their common cause, the two groups reunited in 1890 as the National American Woman Suffrage Association. Another antebellum reform, temperance, resurfaced. Closely linked to suffrage, temperance crusaders became national Woman's Christian Temperance Union (WCTU) members. Organized in 1874 and led by Frances Willard, the WCTU ultimately attracted a broad spectrum of women, who, in addition to closing thousands of saloons, spread their reform efforts into other areas. For example, they established the Young Women's Christian Association (YWCA) in Boston in 1867.

Yet another offshoot of humanitarian reform sprang from Clara Barton's relief work during the Civil War. Barton believed that not only the military should have medical care and supplies, but natural disaster victims should as well. She established the American Association of the Red Cross in 1881, and the organization was reincorporated in 1893 as the American National Red Cross.

Focused on intellectual stimulation rather than specific reform, the women's club movement started in 1868. The idea quickly spread, and women across the country united in clubs for personal enrichment and female fellowship. The General Federation of Women's Clubs, organized in 1892, saw its membership reaching 150,000 by century's end. A similar group, the Association of Collegiate Alumnae, founded in 1882, provided an intellectual community for college-trained women.

In addition to association building, many young, single women joined the white-collar labor force as clerical workers. Introduction of the typewriter in the 1870s guaranteed that routine jobs as typists, stenographers, and telephone operators would fall to lower-paid women. Concurrently, the advent of department stores created a need for retail clerks, and women moved into these positions. By the turn of the century, office and sales clerks jobs were women's work. Factory labor increasingly became the lot of immigrants, who, along with blacks, also filled most of the domestic-servant jobs. For these working-class women, the Knights of Labor sponsored about 270 ladies' locals and another 130 mixed local groups.

As more women worked outside their homes, others acquired higher education. By 1900 about eighty-five thousand women were attending colleges. Trained beyond available professional opportunities, educated women moved into social work, elementary school teaching, and nursing. The nurturing fields had always been women's work, and now these fields were becoming professionalized.

Social work became a profession in 1889, when Jane Addams and Ellen Gates Starr launched the settlement house movement by establishing Hull House in Chicago. The concept spread, and by 1900 there were about one hundred such houses across the nation. Similarly, as public schools mushroomed after the Civil War, demand for elementary school teachers grew apace. States opened separate teacher training schools for whites and blacks, and women filled a customary role of instructing the young.

There was one woman, however, who argued against societal restrictions. Charlotte Perkins Gilman, in her 1898 book *Women and Economics,* maintained that women should be economically independent rather than in bondage to men. In thought and writings, she presaged twentieth-century feminism, declaring that young boys and girls should not be locked into the stereotyped sex roles of the dominant breadwinner male and the subordinate domestic-servant female.

Still another field that became professionalized during this era was nursing. The Civil War had revealed the advantages of females as nurses in the military and in the sanitary commissions, and even physicians, who had been skeptical of women's value, soon advocated professional training. In 1868, physician Samuel D. Gross, president of the American Medical Association, called for training institutions in all major towns and cities. Three years later *Godey's Lady's Book*, a widely read magazine, popularized the idea of "lady nurses."

Several training schools had already begun. In 1857 physicians Elizabeth Blackwell and Marie E. Zakrzewska started training a few nurses for six months at their New York Infirmary for Women and Children. Four years later the Women's Hospital of Philadelphia, run by two women physicians, Ann Preston and Emmelin Horton Cleveland, instigated a six-month practical training course for nurses. In 1863 the New England Hospital for Women and Children, founded by Dr. Zakrzewska, offered a similar program.

After the New England Hospital for Women and Children got a new building in 1872, it launched the country's first general training school for nurses. Female physicians gave weekly lectures on medical or surgical subjects, and students received practical training in medical, surgical, obstetrical, and night nursing under the supervision of other nurses. After a year's study, one of the five students, thirty-two-year-old Linda Richards, earned a diploma, becoming the first trained nurse in the United States. The first black nurse to graduate was Mary Eliza Mahoney, who finished the course in 1879. The hospital expanded its program to two years in 1893 and to three years in 1901.

Three other training schools, established in 1873, had two-year programs and were patterned after Florence Nightingale's work at St. Thomas's Hospital in London, with the superintendent being a nurse. One training school opened at Bellevue Hospital in New York, which began a class with five students under the direction of Sister Helen of All Saints Sisterhood. Next came the Connecticut Training School at the New Haven Hospital, where four aspiring nurses were under the charge of a Miss Bayard. Soon the Boston Training School at Massachusetts General Hospital operated a similar program, with four students under the superintendent, Mrs. Billings. Not surprisingly, many of the early pioneers in establishing professional training programs for nurses had themselves been nurses with the Army or had worked in the U.S. Sanitary Commission during the Civil War. Trained nurses quickly proved their worth in hospitals, and the idea of nursing schools grew. By 1880, fifteen schools instructed 323 students and had 157 graduates. A few males entered the nursing field, and in 1888 Bellevue Hospital in New York City opened the Mills Training School for Men. Fifteen entered the first class; four graduated. The next year forty-four men began the training, and eighteen graduated.

By the turn of the century, the general hospital movement had brought a proliferation of facilities, and the public's concept of hospitals had changed. No longer places of inevitable death, hospitals were havens of healing. More patients created a demand for more nurses, and by 1900 there were 432 hospital training schools with more than 11,000 students. Most training courses had increased to three years. Not only hospitals ran schools, but so, too, did Catholic religious orders. Separate schools, beginning with the Spelman Seminary in Atlanta, trained black nurses. The first university training class for nurses began in 1899 with a course called "hospital economics" at the Teachers' College, Columbia University.

As nursing became a recognized profession, educators clamored for uniform standards of instruction. In 1893 Isabel Hampton Robb took the lead in establishing the American Society of Superintendents of Training Schools for Nurses to set up universal training criteria. Three years later Robb founded the Associated Alumnae of the United States and Canada for all trained nurses. In 1900 Robb, along with other nursing professionals, launched the *American Journal of Nursing*, a periodical for and by nurses.

At the same time these organizational steps occurred, two other trained nurses, Lillian D. Wald and Mary Brewster, took the settlement house philosophy to New York City's Lower East Side. They opened the Henry Street Settlement, providing nursing care and advice either at the settlement house or in the homes of the immigrant poor. In effect, they began public health nursing.

By the end of the nineteenth century, trained professional nurses had earned the respect and acceptance of doctors, hospitals, and the general public. They had broken out of the restrictive cult of domesticity by going to training schools and working outside their homes, but their duties of healing, nurturing, and caring for the infirm reflected the prevailing social norms for women. Nursing was women's work.

The Spanish-American War, 1898

The United States would soon need some of these trained nurses as the nation raced toward war with Spain in 1898. Influenced by European colonialism, the country embraced the lure of foreign markets, the cult of Anglo-Saxon superiority, a sense of national mission, big-Navy enthusiasm, and jingoism. The United States soon liberated Cubans, Filipinos, and Puerto Ricans from oppressive Spanish rule and, in turn, relieved Spain of her overseas possessions.

It took little to generate war fever. In February 1898 an indiscreet letter written by Spanish minister Dupuy de Lome described President William McKinley as "weak"; the letter fell into the hands of the press. A few days later, on the fifteenth, the battleship *Maine* exploded in Havana Harbor with the loss of 266 men. A subsequent investigation and failed negotiations with Spain fanned American anger. Amid cries of "Remember the *Maine!*" Congress declared war on 21 April, and the next day a naval force blockaded Cuba. The Navy's performance during the hostilities would be decisive.

On a second front the Navy dispatched Commo. George Dewey with the Asiatic Squadron to Manila Bay in the Philippine Islands. Dewey reduced the dilapidated defending Spanish squadron to scrap without delay. His ships remained in Manila Bay until U.S. Army troops arrived, captured Manila on 13 August, and occupied the Philippines. Meanwhile, Rear Adm. William T. Sampson led a squadron to Cuba's Santiago Harbor and on 3 July annihilated another major squadron of the Spanish fleet. About sixteen thousand Army troops, carried in naval transports, had landed on Cuba's southeastern coast. These forces sustained heavy casualties at El Caney and San Juan Hill but finally captured Santiago on 14 July. Other troops took Puerto Rico, and the most popular of all American wars ended on 12 August 1898.

In December the peace treaty gave Cubans their independence and ceded Puerto Rico, Guam, and the Philippines to the United States for $20 million. That same year, the country annexed the Hawaiian Islands, Samoa, and Wake Island. The "splendid little war," as Secretary of State John Hay called it, produced an empire for the United States, an empire with overseas possessions and colonies, subjects, and protectorates.

But the price of liberating or acquiring these possessions had been high in human terms. When the Spanish-American War began, the Army increased from 28,000 to about 200,000 troops. Lacking medical facilities and personnel to care for this large influx, the War Department accepted the offer of the National Society of the Daughters of the American Revolution (DAR) to screen potential military nurses. Physician Anita Newcomb McGee directed this examining process. Unlike the Civil War nurses, Army nurses had to be training school graduates, and eventually the service utilized 1,563 contract nurses, paying each thirty dollars a month and subsistence. Other contract nurses came into Army service through the American National Red Cross, which kept a nationwide list of trained nurses. Clara Barton had offered the agency's services to the armed forces when the war began, and about 700 Red Cross nurses helped in Army hospitals.

These nurses served in the United States, Cuba, Puerto Rico, Hawaii, and the Philippines, and on three ships, including hospital ship *Relief.* The Army had acquired *Relief,* a passenger liner, and used the vessel to evacuate casualties from Cuba. Six trained nurses, including Esther V. Hasson, who would later be Navy Nurse Corps superintendent, cared for 1,485 sick and wounded men aboard *Relief.* These infirm were only a fraction of the Army's casualties: 968 deaths from battle, 5,438 from disease. Ravaged by typhoid, malaria, yellow fever, and dysentery, fully 30 percent of American troops were sick by August 1898.

The Navy as well had its share of casualties. Adding 128 ships and raising the complement of officers and men to 26,102 increased the chances for injury and disease. During the war the Navy sustained 85 deaths: 29 from wounds or injuries, 56 from illness. Fourteen Marines succumbed to an outbreak of yellow fever at Key West, but the naval services did not suffer the severe epidemics that had decimated the Army. To help care for the expanded naval force, the Medical Department took on sixty-one volunteer assistant surgeons. One of these, John Blair Gibbs, was killed at Guantanamo, the only naval medical officer to die in the war.

The sick and wounded required nursing attention as well as physicians' services, and the Navy finally got its Hospital Corps in June 1898. Designed to provide a trained cadre of men willing to make a career in the naval service, the Hospital Corps soon sent "trained" male nurses and apprentices to all naval hospitals. Their training lasted only several months and was not comparable to the two to three years women spent in civilian hospital training schools. The network of naval hospitals now included facilities at Portsmouth, New Hampshire; Chelsea; Brooklyn; Philadelphia; Washington; Norfolk; Pensacola; Mare Island; Yokohama, Japan; Newport; and Sitka, Alaska.

Although the Hospital Corps provided "nurses," the Navy still needed professionally trained graduate nurses. Unlike the Army, however, the Navy had no authority to hire contract nurses and so turned to volunteers. Four female students from Johns Hopkins Medical School served as nurses at the naval hospital in Brooklyn. Six women from the DAR register of trained nurses volunteered for duty at Norfolk Naval Hospital and were there for an average of fifty days. Five Sisters of Charity from St. Vincent's Hospital in Norfolk joined in the work. The Navy gave them lodging, board, and transportation, but their pay came from private funds. Later, Surgeon General William K. Van Reypen described the nurses' performance as thorough and conscientious. When the hospital at Portsmouth, New Hampshire, added two new pavilions for housing Spanish prisoners, the Red Cross provided one hundred cots and six trained nurses. These twenty-one women were the first to serve as trained nurses in U.S. Navy hospitals.

The Bureau of Medicine and Surgery had long advocated a hospital ship to accompany fleets engaged in action, and on 7 April 1898 the Navy bought *Creole*, a new steamer operated by Cromwell Lines. Converted to the ambulance (hospital) ship *Solace* and equipped with the most modern medical supplies and conveniences, the vessel could carry two hundred patients and had four surgeons, three hospital stewards, and eight male nurses in her medical department. Throughout the summer she collected the sick and wounded from Cuban operations and transported them to military and naval hospitals in the United States. Among the patients were forty-eight Spanish naval officers and men who were taken to Norfolk Naval Hospital. On her return trips to the Caribbean, *Solace* took medical stores and supplies, as well as assorted "delicacies and comforts" donated by private individuals and societies, to the ships on station there. The *Solace* was the first American ship to meet the Geneva Convention requirements and the first to fly the Geneva cross flag.

By the end of the nineteenth century, the U.S. Navy had developed into a world-class naval power with fleets to sail and colonies to administer. The Navy's evolution had been uneven and sporadic, and growth had depended on external threats or, by the 1890s, on markets to open, colonies to supervise, and primitive people to uplift. Concurrently, training and requirements for officers and men had become professionalized.

As the Navy developed, so, too, did its Medical Department. Beginning with only a few surgeons assigned to ships' sick bays, the department grew into a complex, modern health-care service with a string of well-equipped hospitals and dispensaries staffed by competent doctors. Naval medicine had kept pace with the trends in civilian medicine and employed such innovations as anesthesia, antiseptic surgery, and bacteriology. It had established the Hospital Corps to provide meagerly instructed "nurses" for hospitals and ships.

But the Medical Department still lacked a vital element: professionally trained nurses—those women who had taken two- or three-year courses in hospital training schools. Nursing, along with teaching and social work, was one of the few professions open to women. During the nineteenth century, many women emerged from their prescribed sphere and went into reform movements, clubs and associations, and higher education. And they were no strangers to the seafaring life; they had been in virtually all types of vessels, both private and naval. The Navy needed professional nurses, and such women had proven their worth during the Spanish-American War. Could the two, in fact, be mutually beneficial?

"I Was a Yeomanette"

2

Lou MacPherson Guthrie

U.S. Naval Institute *Proceedings*
(December 1984): 57–64

WHEN THE RURAL SCHOOL in which I'd been teaching near Charlotte, North Carolina, closed in the spring of 1917, I took a civil service examination for clerical work. Since the tests were based on about seventh-grade level, which I'd been teaching, I found them very easy.

I specifically remember a problem on the arithmetic test which asked, "If a bin's dimensions were three × five × seven feet, and it was filled three-fifths full of molasses, how many quarts, or gallons, of molasses would be in it?" Well, I breezed through the multiplication and division and got the number of gallons at once. Then, my mind suddenly blanked. I couldn't remember how many quarts were in a gallon: four or eight. So I guessed eight and handed in my paper.

As I returned to my seat to wait for a friend who'd taken the test with me, I heard muffled laughter at the front desk. I turned and met the eye of the examiner, who motioned me up to his desk.

"Miss," he said, "can you tell us how many quarts there are in a gallon?"

I looked at him miserably and whispered, "Four."

"Right," he said. "That was a tricky question. You are the first person I've seen work it out correctly and quickly. Except for that little mistake, you would have had a perfect paper." He marked it 95 and turned to get the next paper. That was a question I'll never forget.

The following week, I received a telegram from the Bureau of War Risk Insurance offering me a job at the princely salary of $1,000 a year. As I'd been getting $100 a month for a six-month school term, there was great excitement at our house over this.

When I left home for my new job, my mother instructed me to go to the YWCA. But, when I reached Washington, D.C., I found the Y was filled. The newspapers had printed a request for patriotic citizens to open their homes to "the influx of women workers arriving daily for war work." I just missed getting into the home of Secretary the Navy Josephus Daniels, who was from down home. But I did get placed in the home of another member of President Woodrow Wilson's cabinet. When I arrive there, I found that the man had been married three times, and had children from all three marriages under this one roof. So, as soon as possible, I moved out and into the home of another "patriotic citizen." Here, I had kitchen privileges, which was fine, as food was expensive and eateries were always filled.

With me in my new quarters were two girls I'd known in college. Since we three girls were all working in different branches of the government, we had different social contacts. Also, we were each from a different state—Virginia, New Jersey, and North Carolina.

We soon found that hometown boys were constantly passing through Washington or were in one of the several nearby Army posts. Every boy who called for a date wanted to bring a couple of buddies along with him. We went to dances at Camp Meade, Camp Meigs, Fort Myer, Annapolis, and the Y. The gay hellos and goodbyes spoken the same evening to khaki-clad boys excited over the prospect of going abroad and ending the war quickly and bloodlessly by a great show of strength glamorized the romance of the times. This was our first war. There were no apprehensions. Chivalry was still in flower.

We always seemed to be singing little snatches of nonsense around some canteen piano, such as—to the tune of "Keep the Home Fires Burning":

"Keep the old clothes going. Do your bit in sewing.
The boys are far away, the bills come home.
If there's still a lining, thru the old clothes shining,
Turn the old clothes inside out, 'til the boys come home!"

There must have been dozens of parodies of other songs. One parody of the song "Ja-Da" went like this:

"Had-a, had-a, had-a on the porch last night.
Kissed her once. Then I kissed her twice.
Kissed her again, because I thought it was nice.
For I had-a, had-a, had-a on the porch last night!"

I know now that there was an unreal and feverish enchantment in the way we danced and sang and made light mockery at the ugly face of war. Never again would boys seem so glamorous as these with their overseas caps set rakishly awry. We smiled and joked and danced many "Paul Joneses" with them to meet new partners. We knew the spell would be broken at midnight, and we probably would never see each other again. But this made it all the more exciting and, at the same time, sad. At the end of these dances, we sang "Farewell to Thee," and not the "Goodnight Ladies" that we'd been accustomed to back home.

The Bureau of War Risk Insurance was temporarily housed in the old Smithsonian Institution while a Navy building was under construction. Over the 30-foot table where I worked hung the skeleton of a whale. It always seemed ready to fall on the heads of the many girls beneath doing desk work. I quickly made friends with Julie from Mississippi, Louise from Virginia, and Lalla from Kentucky. All were beautiful girls in their late teens.

Many funny letters came into our department. I entered some of the great lines in my diary. The more entertaining ones included:

"Just a line to let you know I am a widow and four children."

"I've been in bed two years with one doctor, and intend to try another."

"Caring to my condition, I have a broke leg number 975."

"Date of birth: not yet but soon."

"I didn't know my husband had a middle name called 'none'."

"Both sides of our parents are old and poor."

"I have a seven months old baby, and she is a girl and can't work."

"If you don't send my allotment, I'll be forced to lead an immortal life."

"Has Bill put in an application for a wife and child?"

"You ask for my allotment number. I have four boys and two girls."

"Please correct my name, for I could not go by a consumed name."

And there was one long, tear-stained letter from an 18-year-old hysterical wife, who'd been so excited when she got her first check that she had to go to the bathroom. There, she accidently dropped it in the john and flushed it away!

Then, there came the wildly exciting day when the U.S. Navy called for 100 of the women war workers to enlist and do accounting work at the Navy Yard. This would release sailors to man the warships. *The Washington Post* and other papers ran many letters violently opposing such business. Wrote one irate retired colonel, "Preposterous! First the women wanted to vote. Then Alice Roosevelt started them smoking cigarettes! Now they're talking about being soldiers. Next thing we know they'll be cutting off their hair and wearing pants!"

But for all the derision and joking about the matter, women were permitted to enlist. Julia, Louise, Lalla, and I enlisted. This examination was easier than the civil service one. The main requirements were to write legibly, make clear figures, spell correctly, and be able to add long strings of figures quickly and without mistake. This was my dish, and I ate it up. I placed among the top five of the 100 women accepted from several hundred applicants.

I could scarcely sleep at night, hugging the thrilling thought of being enlisted in the Navy. "Where would they send me? Would I be assigned to a transport ship or maybe to some foreign port?" All four of us girls placed high on the test, so we thought we'd probably stay together.

The morning that my mail brought my assignment to the USS *Triton* was terrific. The four of us were given similar assignments, as were the remainder of the 100 girls. But I soon learned that the *Triton* was an old, worn-out tug that would have probably sunk to the bottom of Chesapeake Bay if we'd all tried to board her. But, technically, we had to be given a ship assignment. It was the Navy yard for us.

Still, we were in the Navy. Our uniforms were blue serge suits. The skirts were ankle length, and the smart ensign-type jackets had naval insignia on the sleeves. We had straw sailor hats for summer and hard, heavy dark blue ones with bands reading "U.S. Navy" across the front for winter. Our navy blue capes looked like those George Washington wore when he crossed the Delaware River. Mine is still, after 65 years, in perfect condition. When we got our uniforms, we all went downtown and had our pictures made to send home.

On Easter weekend, Julia, Louise, Lalla, and I decided to go to New York—the first visit to the city for all of us. We rode the train all Saturday night and arrived at daylight Sunday morning. After a hearty breakfast at the nearest lunchroom, we set out on a sightseeing trip via street cars. We took in Chinatown, the Bowery, and Greenwich Village before ten o'clock. When we went back uptown, I thought I had never seen so many friendly people. They were interested in our uniforms. And they were interested in us. They all smiled at us as we passed by. It made me feel proud, and I marched more erectly.

None of us had ever been in a bar, so we went into the Hotel McAlpine, where there was a delightful bar facing the street. Since we didn't know what any of the drinks were, we each decided to order a different one. The only two drinks that I recall were the one which Louise ordered called "Horse's Neck," which turned out to be beer, and the one I ordered.

When the waiter brought the tray of drinks, he asked, "Which one is the Sport?" At our blank looks, he asked, "Who ordered the Scotch?"

"Me," I said brightly. "My people all came from Scotland, so I wanted to see what the Scotch people drink." He looked at me a little strangely and handed me a small glass of liquid—not at all like the tall, creamy ones with cherries that Julie and Lalla had gotten. But anyway, I tossed it down in a gulp. And then drank glass after glass of water to put out the fire.

But it was a gay time. We laughed at everything. I remember at one point during the morning, Julia asked the others, "Shall we tell her now?"

"Let's tell her," Lalla snickered.

"Tell who, what?" I asked.

"Look at your heel," Louise ordered.

I looked down and gasped. Too horrified to speak, I merely gazed at my regulation black stocking. Its heel was entirely ripped out, and white skin shone from beneath the long blue skirt. There it was, Sunday morning, and all stores were closed. There was no needle or thread available. Fortunately, Louise had her fountain pen with her, and, fortunately, it was filled with black ink. So I blackened my heel and went merrily along.

When we went to work Monday, we found that a swingshift system had been installed. We were to finish out this month at 8–4. Then, we'd have 4–12 for two months, and 12–8 for the next two. Thus, the amount of work done at the cramped facilities of the Navy Yard would be tripled until the building at 17th and B was completed.

We liked the midnight shift best, even though we had to get off the street car at midnight in the worst section of the city and walk down the wharf. Here, wharf rats nearly as big as opossums scuttled across our path in the moonlight. But it was quite safe. There was very little crime recorded in the city then. We felt no timidity about walking alone at midnight on poorly lighted streets.

We had a chow break at 2:30 A.M. when our office was closed for ventilation and cleanup. But we four usually felt too sleepy to eat. Each night, we'd roll two of the huge flattop desks together at the back of the room and curl up and fall asleep like a nest of kittens.

This was quite aggravating to the 17-year-old sailor who swept and straightened the room. He fussed, but we ignored him. So one night, he filled two wastebaskets full of paper and put them on each side of our "bunk." Then, he set fire to them. The smoke didn't even wake us from our slumber, but it did bring the wrath of the higher-ups. They put him in the brig, and he almost got a dishonorable discharge.

After we got off duty at 8 A.M., I'd often stop for breakfast at the Union train station dining room while waiting to transfer to another streetcar. Once, I recall being very hungry. I sat down at the only empty table, which had just been hurriedly vacated by a man whose train was called. The busy waitress ignored me, so I ate the remains of his scarcely touched breakfast.

I remember one lunch hour when I went to the Navy dentist to get my teeth cleaned and examined. This was after we'd moved into the new building at 17th and B. The young and personable dentist was alone in the clinic. I was embarrassed to be sitting there with my mouth gaping open so this handsome man could do the dental work. When he had finished, without saying a word, he suddenly pinned me in the chair and gave me one of the biggest, longest kisses I've ever had. Then, just as suddenly, he released me. I fled from the office. When I told the other three girls what happened, they all immediately made appointments to get their teeth cleaned and examined. But not one of them got the treatment I did. Perhaps the dentist, like me, was a misplaced hayseed. Or maybe I reminded him of the girl back home.

By now, many more girls were enlisted in the Navy. The work was becoming routine. The glamor was gone. The boys were overseas. We felt stalemated in Washington. Worse still, a rumor started that we were being moved across the river next to a fertilizer factory in Alexandria.

But then, suddenly, the war was over! Armistice Day was wildly exciting. All offices closed. We danced in the streets with any passerby. Everyone was happy. Everyone laughed and celebrated. On 12 November, it was hard to get back to our humdrum duties. But then began a series of parades to greet military brass and returning units. And there were parades of any old kind. The yeomanettes were always in the vanguard.

The civil service girls were now returning home in droves. But it took string-pulling, personal friendships, and hard work to get released from the Navy. But, at last, I came home to Charlotte for a few months rest. Then, I returned to a civil service job, working in a veteran's hospital in Asheville, North Carolina. But I only worked there for six months.

By the next fall, I was back in the schoolroom teaching seventh-grade arithmetic. And helping the students figure out how many gallons and quarts it would take to fill a three × five × seven-foot bin three-fifths full of molasses.

Mrs. Guthrie earned a bachelor's degree in journalism from University of North Carolina, Chapel Hill. She has worked as a teacher and a principal in schools in Charlotte, North Carolina.

3 "Womanpower in World War I"

Susan H. Godson

U.S. Naval Institute *Proceedings*
(December 1984): 60–64

AFTER WORLD WAR I had been raging for nearly three years, Secretary of the Navy Josephus Daniels foresaw an acute manpower shortage if the United States became involved. In early 1917, he asked his legal advisers if any law prohibited women yeomen. The startled lawyers investigated and found that the section of the Naval Reserve Act of 1916 establishing the Naval Coast Defense Reserve specified "persons" rather than "males."

"Good," said Daniels. "Enroll women in the Naval Reserve as yeomen, and we will have the best clerical assistance the country can provide."

In March 1917, the Bureau of Navigation, charged with naval recruiting, authorized the commandants of naval districts to enlist women as yeomen, radio electricians, or other useful ratings. The commandants were shocked by the innovation at first, but they ordered nearby recruiting stations to start signing up women. Within a month, 200 eager young women had become yeomen, and their numbers increased rapidly to a peak strength of 11,275. No centralized administration or separate organization for women emerged. They were simply a pool from which the Navy could draw needed help.

To join the Navy, a woman went to a recruiting station, was interviewed for qualifications, and filled out application forms. A typist, stenographer, or translator took a brief test. If she passed the physical examination, the woman then swore to support and defend the Constitution of the United States. The entire process often took less than a day. The new recruit had speedily signed up for a four-year hitch as a yeoman in the U.S. Navy. The "F" for female appeared a little later, after a number of women had mistakenly received orders to combatant ships.

Requirements for service were few. The women had to be between 18 and 35 years old. One 15-year-old did sneak in, however. She enlisted along with her mother. The women had to be of good character and neat appearance and often had a person of known integrity sponsor them. Preferably, they were high school graduates with business or office experience. There were no specified higher educational standards.

Although the yeomen (F)—sometimes called yeomanettes—volunteered from every part of the country and from Alaska, Hawaii, and Puerto Rico, more women came from Massachusetts, New York, Pennsylvania, the District of Columbia, and Virginia than from other states. They came from every walk of life, every socioeconomic background. Some were sheltered daughters of wealthy parents; others were daughters of cabinet officers or relatives of congressmen or high-ranking naval officers. Most came from middle-class homes.

Naval service was often a family affair. Two sisters serving side by side was commonplace. But more unusual were two Massachusetts families who each sent four daughters to become yeomen (F). Although some women found Navy pay, generally higher than civilian wages, enticing, one joined the Navy because of the "super" uniforms. On the whole, however, yeomen (F) had a sincere, patriotic desire to help the war effort.

The Navy provided no formal indoctrination for women recruits. They started working immediately in offices around the country. As they coped with the strange nautical vocabulary and learned their duties on the job, many attended training classes after work. Since they were mostly clericals, enlisted women had to acquire yeomen's skills. For three months, the yeomen (F) went to night schools to learn naval routine and procedure. They mastered regulations about ratings, disratings, transfers, enlistments, discharges, and service records. They learned about naval correspondence and had to type 21 error-free words a minute. They became experts in all types of naval regulations. To rise to chief yeomen, they also needed a firm knowledge of bookkeeping.

Drills supplemented schools as a part of training. Anticipating a quick triumph, the Navy wanted women ready to participate in victory parades. For 30 minutes one day a week, women practiced marching in straight columns. Some carried rifles for added military impressiveness. Manipulating their long skirts while marching in tight lines and following unhousebroken horses was a major problem for the women sailors.

As women adjusted to life in the U.S. Navy, formerly an all-male bastion, they also had to change their living habits. Although some commuted from their homes, most had to find a place to stay. Housing was scarce in wartime America, and the Navy provided no barracks or mess halls for women. They rented rooms or shared apartments or houses. Some were fortunate enough to live with family friends. The Navy paid yeomen (F) $1.25, then $1.50, a day in subsistence allowances.

At first, yeomen (F) wore civilian clothes with arm bands indicating their ratings. No one had devised a uniform for them. In 1918, the Navy apparently decided to cover any trace of feminine curves and issued a severe uniform of blue serge for winter and white drill for summer.

Secretary of the Navy Daniels established a startling compensation policy. "A woman who works as well as a man ought to receive the same pay," declared Daniels. The Navy carried out that policy. Like their male counterparts in the regular Navy, chief yeomen earned $60 a month; yeomen 1st class, $40; 2nd class, $35; 3rd class, $30, plus their subsistence. They also received a uniform allowance, medical care, and war risk insurance.

For confirmation in ratings, yeomen (F) had to serve at least three months in those ratings. If they performed well, they could move higher. After passing a written examination and securing the recommendation of their commanding officer, women were eligible for promotion. Chief yeoman (F) was their highest ranking. Although many rumors circulated about commissioning women, especially as ensigns in the Pay Corps, few women ever exercised administrative responsibility. There were no women officers in World War I.

Governed by the same disciplinary regulations as men, yeomen (F) learned to obey orders and follow routine without question. Sometimes, they erred and lost liberty or pay as punishment. Daniels refused to subject women to court-martial because he thought it contravened public policy. He disapproved the courts-martial of several yeomen (F).

As more and more women joined the Navy, they served in the bureaus in Washington, at Navy yards, in all naval districts, and many shore stations. They often a worked ten hours a day, six days a week; they expected no preferential treatment because of their sex.

Their duties were primarily clerical. Most were stenographers, typists, clerks, accountants, and bookkeepers. Not all clerical jobs were routine, however. One yeoman (F), for example, chronicled the performance of the Navy's 1,500 pay officers, and another directed priority orders sent to railroad officials to move the Navy's supplies. One auditor regularly signed multimillion-dollar vouchers, and a yeoman (F) oversaw manufacturers producing clothing for 250,000 sailors. One secretary handled all correspondence involving the Navy's armed guards on merchant ships. She received the first news about ship sinkings and about armed guards captured by the Germans, and she had to notify relatives of casualties. Other yeomen (F) were messengers, entrusted with carrying secret offices documents to scattered naval offices.

Yeomen (F) rapidly became the, Navy's switchboard operators, and they handled calls speedily and efficiently. Because of their outstanding performance, women became the permanent operators of naval phone systems. In other forms of communication, some yeomen (F) operated radios and telegraph keys or broke codes and decoded cables in the United States and in Puerto Rico. A number of them translated foreign documents and newspapers.

Some women were commissary stewards, and others were librarians at naval training centers. Several became the Navy's fingerprint experts, responsible for all naval fingerprinting. Others were draftsmen or camouflage designers. Many hospitals had yeomen (F) on their business staffs. Five women in the Bureau of Medicine and Surgery went to France with naval hospital units. At the Naval Medical School, a yeoman (F) divided her time between helping prepare a training course for new naval physicians and standing by the phone in case the White House needed to reach the Surgeon General. One woman worked with the Office of Naval Intelligence in Puerto Rico. Others served in Hawaii.

Because of the manual dexterity attributed to women, many overall-clad yeomen (F) spent their enlistments on production lines. They assembled delicate torpedo parts at the Newport Torpedo Station, Rhode Island, put together primers at the munitions factory in Bloomfield, New Jersey, and assembled weapons at the gun factory in Washington, D.C.

Highly visible because of their uniqueness and their unusual uniforms, yeomen (F) helped with Liberty Bond drives and rallies and served at recruiting stations around the country. One woman recruiter in New York City signed more than 10,000 men for the Army and Navy and received a gold medal from the American Patriotic Society.

Yeomen (F) marched in many parades. They participated in Victory Loan drive parades and in the New York City parade of 22 February 1919, celebrating both President George Washington's birthday and President Woodrow Wilson's return from treaty negotiations in France. They formed the honor guard at Union Station when President Wilson arrived at the nation's capital. When the famous 42nd Rainbow Division returned from France in triumph, a yeoman (F) battalion smartly marched down Pennsylvania Avenue in the parade.

Shortly before the war ended, the Marine Corps also began enrolling women. The Commandant, Major General George Barnett, believed that 40% of the Marine Corps' clerical work could be done by women, although three women, he said, were needed to do the work of two men. Receiving the go-ahead from Secretary of the Navy Daniels, the Marines began recruiting women in August 1918. Only 376 Marine Reserve (F) were on active duty when the war ended, and they performed duties similar to those of their sisters in the Navy.

During the devastating influenza epidemic of 1918, yeomen (F) and women Marines volunteered to help care for the sick. At first, the Navy suggested the women wear gauze masks to prevent the spread of the disease. When the masks proved ineffective, the Navy recommended they have a daily shot of straight whiskey, chased by black coffee, to ward off germs. "I learned to drink straight whiskey during working hours in the U.S. Navy," laughed one ex-yeoman (F). Despite these precautions, about 37 women died while on active duty, many from influenza.

When World War I ended, demobilization of the armed forces began. Slowly, in late 1918 and the first half of 1919, the Navy released yeomen (F) from active duty. By midyear, nearly all women had returned to civilian life. They did not plan on a permanent career in the military. The women stayed on inactive status until their four-year enlistments expired. They received a retainer fee of $1 a month. Eligible women received the Good Conduct Medal and the World War I Victory Medal when discharged from the inactive reserves. They could join the newly formed American Legion, and they got veterans' preference for civil service jobs.

After they left the Navy, most yeomen (F) returned home. But others took government positions. Thousands joined service leagues and the American Legion and even organized their own Legion posts. Several hundred became WAVES, women Marines, or WACS in World War II. They remembered their service in World War I with pride and enthusiasm. "We had a small part in the great Allied victory," said one. "I loved it and felt like I was doing my bit," remembered a former chief yeoman

(F). "We didn't realize it at the time, but we were trailblazers for women in the military," recalled another. "I wouldn't give up one single minute of my service," emphasized one woman. A young widow of a serviceman thought her enlistment inspired other women. She was "happy in doing my duty for the country."

Always their greatest admirer and strongest advocate, Secretary Daniels called the yeomen (F) "the elect of their sex." When the women left the Navy in July 1919, he stated: "It is with deep gratitude for the splendid service rendered by yeomen (F) during our national emergency that I convey to them the sincere appreciation of the Navy Department for their patriotic cooperation." Rear Admiral Samuel McGowan, Paymaster General of the Navy, exemplified the thoughts of career naval officers: "The war efficiency of the Navy Department is due, in a big part, to the excellent work of the women employed in it. . . . The women who have men's jobs in my department have shown themselves as efficient as men."

Appreciation of the yeomen (F) seemed short-lived, however. In 1924, Congress debated a bill giving bonuses to veterans. Some congressmen wanted to exclude women from eligibility, but, prodded by former yeomen (F), the American Legion lobbied actively. Other service organizations and women's groups joined in demonstrating the injustice of the exclusion. When Congress passed the Adjusted Compensation Act in May 1924, Yeomen (F) and women Marines shared in the bonuses.

The yeomen (F) of World War I were pioneers. As the first enlisted women of any service, they released desk-bound men for combat. Their dedicated wartime service temporarily opened different occupational opportunities for women and ushered in a beginning of equal pay for equal work. The yeomen (F) were forerunners of a fuller involvement of women in American life.

"The Waves in World War II"

4

Susan H. Godson

U.S. Naval Institute *Proceedings*
(December 1981): 46–51

WOMEN'S SERVICE in the Navy is not a new idea; it began in 1908 with the Navy Nurse Corps. Later, as U.S. entry into World War I drew near, Secretary of the Navy Josephus Daniels authorized enrolling women in the Naval Reserve. When the war ended, 11,275 women had become yeomen (F), the "F" for "female." By filling shore positions in Washington and naval districts, they replaced men needed for sea duty. There were no women officers, and most yeomen (F) were clerical workers. The women returned to civilian life in mid-1919.

Late in 1941, when U.S. involvement in a two ocean war presaged a severe manpower shortage, military planners once again considered allowing females in the armed services. After Congresswoman Edith Nourse Rogers (Repub.-Mass.) sponsored legislation establishing the Women's Army Auxiliary Corps, she asked Rear Admiral Chester W. Nimitz—whose Bureau of Navigation was responsible for naval personnel—whether the Navy would like a similar bill. Nimitz was lukewarm to the idea. He nevertheless asked various bureaus and offices if they could use women as substitutes for men in shore jobs. The only favorable responses came from the office of the Chief of Naval Operations and the Bureau of Aeronautics. As naval leaders bickered for several months, one woman observer quipped, "Many admirals would prefer to enroll monkeys, dogs, or ducks."

In April 1942, Rear Admiral Randall Jacobs, the new chief of the Bureau of Navigation, requested all shore and headquarters stations to send lists of positions that women could fill. Once again, most responses were negative. The greatest enthusiasm came from the Bureau of Aeronautics, the office of the Chief of Naval Operations, and naval communications and intelligence. As public and congressional interest grew, Jacobs began to list potential jobs for women and sell the idea to naval leaders.

Realizing that Congress would pass some type of legislation for a women's reserve, the Navy appointed Elizabeth Reynard, a professor of English at Barnard College, to help devise a program for women. The energetic, British-born Reynard worked as a special assistant to Admiral Jacobs. The press had begun calling the future reservists "sailorettes," "goblettes," and "swans," so one of Reynard's early assignments was to choose a nickname for the women. She suggested the nautical sounding Waves—Women Accepted for Volunteer Emergency Service.

Needing more guidance on organizing the Waves, the Navy again looked to the educational world and appointed Dean Virginia Gildersleeve of Columbia University to head an advisory council. The council, which began meeting in April, consisted of prominent women educators throughout the country. Especially useful in the formative months of the Waves, the council worked with the Navy in setting up standards and procedures for the new group.

An important early task of the advisory council was to recommend a director for the Waves. Hoping to attract qualified applicants to the service and also to reassure the women's families, the council suggested Mildred McAfee, president of Wellesley College. She had a great deal of experience in working with young women and accepted the challenge of supervising the military program.

While the Navy made its tentative plans, Representative Melvin Maas (Repub.-Minn.) introduced a bill in March 1942 to provide for a women's reserve as part of the Naval Reserve. In the Senate, Naval Affairs Committee Chairman David I. Walsh (Demo.-Mass.) adamantly opposed women's entry into the Navy. Such a move, he argued, would lead to the breakup of American homes and eventually to the decline of civilization. Later, however, he reluctantly agreed to a women's auxiliary for the Navy similar to the Army's.

If the Navy had to have women members, it wanted tight control over them for reasons of security, assignments, and convenience. A loosely managed auxiliary that was not an integral part of the service was unacceptable. Thoroughly alarmed that Congress would pass a bill without its approval, the Navy worked feverishly to get its ideas accepted. Finally, after presidential intervention, Congress approved having the women's group *in* the Navy and not an auxiliary *with* it.

The Women's Reserve of the U.S. Naval Reserve was established on 30 June 1942. Women were subject to military discipline, security regulations, and placement, and could serve until six months after the war ended. As in 1918, the main purpose of the female reservists was to release men for duty at sea. Initially projecting an enrollment of 10,000 officers and enlisted women, the new law restricted their highest rank to one lieutenant commander, allowed for 35 lieutenants, and prohibited service outside the continental United States or on board naval vessels and combatant aircraft.

Director McAfee received the rank of lieutenant commander and was sworn in as "an officer and gentleman in the United States Navy." Quickly sensing the ill-disguised hostility of many toward service women, McAfee jokingly noted the similarity between naval men and the author of the 88th Psalm: "Thy wrath lieth hard upon me, and thou hast afflicted me with all thy Waves." Moving rapidly to get the Wave program under way, McAfee operated the Women's Reserve office in the Bureau of Personnel, formerly the Bureau of Navigation. Her five-member staff consisted of an assistant director, a public relations officer, a traveling representative, and two special assistants.

These women and another dozen officers appointed during the summer of 1942 knew little of naval traditions, practices, or even vocabulary. Nevertheless, they planned recruiting procedures, training programs, and possible assignments for Waves. Drawn from the business and professional worlds, the officers personified respectability and competence. This small group was instrumental in shaping the direction of the Women's Reserve.

To attract more women, the Navy launched effective recruiting and publicity campaigns. Lieutenant Louise Wilde, a public relations officer, diligently worked to stir up patriotism among women and a desire to serve. In addition, regional procurement and recruiting offices used Waves to stimulate more interest in the naval program. Waves gave speeches, posed for pictures, interviewed prospective recruits, marched in parades, and attended ship launchings. Although the publicity emphasized the glamorous aspects of naval service—fashionable uniforms, training on college campuses, unusual work, and opportunities to meet eligible men—Waves rejected tasteless advertising. Cheesecake photos and blatant promises of dates or matrimony would never have convinced prospective Waves, their parents, and certainly not the churches of the careful supervision given the young women.

As a result of the energetic promotion campaign, Waves recruited women from every state in the union and from all socio-economic backgrounds. They were primarily motivated by an intense desire to aid the war effort. Waves had to be at least 20 years old, of high moral character, in good health, and had to meet the Navy's tough standards by passing rigorous verbal, mathematical, and physical examinations. They had to be high school graduates or have completed two years of secondary schooling and worked for two years. Officer training required a college degree or two years of college with two years of work experience.

Following the advice of the advisory council, the Navy used college campuses for training, both because of the dignity of the academic setting and because of the readily available facilities. Smith College at Northampton, Massachusetts, became the site for officer indoctrination. A small group of women underwent a four-week training session in August and September 1942. These officers became administrators and teachers at Women's Reserve schools. In October, the first regular class reported to the "USS *Northampton*," commanded by Captain Herbert W. Underwood, who was recalled from retirement to administer the training program.

Officer candidates made the rapid transition from civilian life to naval routine and discipline. They learned about naval history and organization, ships and aircraft, and communications and law. They shared crowded dormitories, coped with double-decker bunks, and stood in line for meals. And they mastered the new vocabulary; such terms as "ladder," "bulkhead," "hold," "deck," "mess," "head," and "chow" took on new meanings. The women also participated in physical education and close order drill.

When the last class of Waves completed its course in December 1944, more than 1,000 communicators, many trained at nearby Mount Holyoke College, and over 9,000 general duty officers had received commissions in the Naval Reserve. In addition to Waves, officer candidates for Coast Guard Spars and women marines received their indoctrination at Northampton until mid-1943.

Initially, the Navy believed a combination of boot and specialist training would suffice for enlisted Waves. In October 1942, the Women's Reserve began sending its recruits to Oklahoma A and M at Stillwater for yeoman training, to the University of Indiana at Bloomington for storekeeper instruction, and to the University of Wisconsin at Madison for radio operator training. Several months later, another school for yeomen opened at Iowa State Teachers College at Cedar Falls. By the end of the war, enlisted women received specialized training at 20 colleges, universities, and training stations.

Desiring more thorough indoctrination for enlisted Waves, the Navy took over the entire campus and facilities of Hunter College in the Bronx.

In February 1943, the U.S. Naval Training Center (Women's Reserve), Bronx, was commissioned and promptly dubbed the "USS Hunter." Under Captain William F. Amsden, a destroyer commander fresh from Pacific convoy duty, the women's college was quickly transformed into a military establishment. The training center took in 2,000 women every two weeks for the six-week boot training. Because of the growing demand for Waves, the training period soon dropped to four weeks.

At Hunter College, enlisted women learned the fundamentals of Navy life, underwent physical conditioning, and mastered such intricacies as the proper salute. The women's qualifications, test scores, and preferences helped determine later placement. After completing indoctrination, enlisted Waves went directly to assigned positions or to advanced training schools. By mid-1945, more than 80,000 Waves and 5,000 Spars and women marines had completed boot training at Hunter.

Waves also coped with new housing, dress, and social regulations. Unwilling to assign enlisted women to bases with inadequate housing, the Navy built new barracks or converted old quarters to accommodate females. Renovations included semi-private cubicles for four women, additional laundry facilities, lounges, and remodeled bathrooms. If government housing wasn't available, Waves found local quarters and received an additional subsistence allowance. Finding mixed accommodations at any base unfeasible, the Women's Reserve preferred all enlisted Waves to live either in barracks or in local housing. Regulations for officers, however, were not as strict, and they could choose Navy or civilian quarters.

Early uncertainty about appropriate uniforms generated some bizarre suggestions. Wave director McAfee defeated a proposal for comic-opera stripes and insignia of red, white, and blue. An experimental hat with a perky, turned-up brim filled with water during rainy-day drills. Finally, the well-known couturier, Mainbocher, volunteered to design a stylish dark blue uniform with light blue stripes, adding a white dress uniform for summer. Modifications in clothing took place as the war continued. Women wore slacks or dungarees only when necessary for work or for sports. Stressing neatness and smart appearance, the Women's Reserve tolerated no sloppily attired members.

Restrictive marriage policies hampered early recruiting efforts. Initially, Waves could marry no one in the armed forces, then new rules permitted marriage to men in services other than the Navy. Finally, Waves after completing training, could marry naval officers or enlisted men.

There were also regulations governing pregnancy. Prohibited by law from having dependent children under the age of 18, pregnant Waves, either married or single, were honorably discharged from the service. A later change in policy permitted women whose pregnancies had ended before their resignations to remain in the Waves. Waves received no allowances for dependent children over 18 years old unless the father was dead or the mother was, in fact, the main source of support. Regardless of their physical conditions, husbands were never counted as dependents. These stringent regulations damaged recruiting efforts and caused some Waves to resign.

Because the Waves were a highly select and volunteer group, disciplinary problems were relatively minor, and the discharge rate remained very low. Few Waves were involuntarily released, and most discharges were for unsuitability or inaptitude rather than for more serious offenses. Location often determined permissible conduct. "Behavior unbecoming a lady," drunkenness and disturbance of the peace, might pass unnoticed in a large city but could shatter the service's reputation in a small town.

Initially limited to one lieutenant commander, the Women's Reserve had difficulty attracting highly qualified and badly needed officers such as physicians. By late 1943, the number of Waves far exceeded original estimates, and Congress promoted the director to captain and removed limitations on lower ranks. As a result, the Waves became an attractive alternative to civilian employment.

Women's military authority remained restricted to their own group, however. Although Captain McAfee thought Waves should be given command of both men and women, especially at training schools, instant opposition from male officers ended any such idea. Men feared that giving women a little military authority would provide "an opening wedge" that might expand to all situations.

Despite a pressing need for Waves at overseas stations, the original law prohibited their service abroad. The Navy finally overcame the objections of conservative congressmen who maintained that ladies should "stay on their pedestals" rather than fight a war. In late 1944, a change in the law allowed Waves to serve in Hawaii, Alaska, and at Caribbean bases within the American area. Early the next year, the first group of Waves assigned overseas reached Hawaii, with their numbers quickly growing to over 4,000. Eventually, Waves fanned out to replace men sent to sea.

At first, Waves were concentrated in traditionally female tasks. Waves handled vast quantities of paperwork generated by the wartime Navy. In all bureaus, offices, and shore stations, yeomen served as secretaries, stenographers, file clerks, and receptionists. Waves filled administrative, public relations, and personnel billets—other familiar fields for women. A large group of Waves worked as storekeepers. Involved in disbursing and accounting tasks, these reservists paid naval personnel and kept track of complex expenditures. They sent supplies to fleets and advanced bases and distributed everything from shoe polish to engine parts.

Rapidly taking over fleet post offices, Waves became the Navy's mailmen. They sorted, weighed, stamped, and checked the enormous volume of mail to and from sailors overseas. V-Mail had to be photographed, and Waves handled this chore. Eventually women reservists managed 80% of the mall.

Always associated with the healing arts, women flocked into the Bureau of Medicine and Surgery. Female doctors, dentists, technicians, and medical illustrators served at naval hospitals, dispensaries, and training stations. The 13,000 enlisted women in the Hospital Corps quickly demonstrated their usefulness. They assisted in wards, operating rooms, laboratories, and offices. They also worked as X-ray technicians and physiotherapists.

As their numbers increased, Waves assumed novel roles. In the Bureau of Personnel, for example, Waves helped to distribute welfare and recreation funds for new ships and stations and settle the myriad of problems from ships lost at sea. Some worked with machines and records in order to keep track of naval statistics.

Waves manned the communications network of the office of the Chief of Naval Operations and stations throughout the country. This highly secret duty, often tedious and monotonous, included long hours of sending and receiving coded radio messages. They checked dispatches and operated teletype machines. An unusual means of communication fell to a few Wave seamen. After learning to handle homing pigeons, these women served at lighter-than-air stations along the coasts. Used in naval blimps on antisubmarine patrol, the trained pigeons carried messages to their handlers at air bases during periods of radio silence. Another select group of Waves, chosen for academic excellence and language aptitude, attended the 14-month Japanese language course at Boulder, Colorado, and eventually monitored Japanese radio broadcasts.

The Bureau of Ordnance, an unlikely place for women, drew mathematicians and technicians. Wave officers attended special schools to learn design, manufacture, and use of the Navy's guns, torpedoes, mines, and bombs. Some mastered chemical warfare techniques while others became familiar with aviation gunnery. Trained women became gunnery instructors or assisted men who handled ordnance production. Some wrote abstracts of battle reports comparing naval and enemy ordnance, while others prepared pamphlets on ordnance assembly and repair.

Using Waves for still more uncommon tasks, the Bureau of Yards and Docks assigned women to purchase real estate, work with secret war plans, and devise camouflage techniques. Late in the war, a few Wave engineers worked in the bureau. Women assigned to the Bureau of Ships assisted shipbuilding supervisors and machinery inspectors. Some officers oversaw materials procurement for ship construction; others scheduled construction or helped develop optical and sonar systems.

One of the most unusual assignments in the Potomac River Naval Command was testing airplanes in the wind tunnel at the Washington Navy Yard. Waves used intricate machinery to chart and determine stress resistance of new models of aircraft. At the nearby David Taylor Model Basin, they helped test and evaluate ship models and equipment.

Some Waves worked in port directors' offices and found themselves close to the action of the fighting Navy. They routed ships in the harbors, attended convoy conferences, and checked on vessels that were ready to sail. They assigned anchorages and debriefed naval gun crews from merchant ships.

The Bureau of Aeronautics, the most receptive to women, eagerly took in nearly one-third of all Waves. Its enthusiasm stemmed mainly from Lieutenant Joy Bright Hancock's initiative. She had been the civilian chief of the editorial and research section of the bureau before the war. As the Women's Reserve representative to the bureau, she pushed for expanded and innovative use of Waves in the air arm of the service.

The bureau's enthusiasm spread to the field, and within the air specialty women performed the most unusual work. About 600 Waves filled the glamorous but demanding tasks of control tower operators. At widely dispersed airfields such as Pensacola, Seattle, and Corpus Christi, Waves kept air traffic moving by directing landings and takeoffs. One thousand Waves taught naval pilots instrument flying in Link trainers, which simulated small cockpits. Parachute rigging—the care and packing of the chutes—occupied other Waves. Realizing that the parachute was a man's last chance, these women diligently mastered the precise demands of rigging. Aviation machinist's mates, the grease monkeys of naval air, were often women reservists. They took apart, repaired, reassembled, and calibrated delicate aircraft instruments and overhauled plane engines. Late in the war, a few Waves, trained as navigators, flew in transports in the United States and to Hawaii and the Aleutians.

The multiple tasks performed by women in the Navy represented new directions in the use of female talents. By the end of the war, the demonstrated ability of the Waves, combined with the manpower shortage, opened 38 of the 62 enlisted ratings to women reservists. At peak strength, 86,000 dedicated officers and enlisted Waves filled positions at over 900 shore stations in the United States and at a few locations overseas.

Waves surmounted much opposition to their presence in "this man's Navy." Fleet Admiral Chester Nimitz, chary about the use of women early in the war, later commended the Waves for their competence, energy, and loyalty. Waves' work, Secretary of the Navy James Forrestal noted, had been overwhelmingly successful and in the highest tradition of the naval service. Similarly, Fleet Admiral Ernest J. King, Chief of Naval Operations, praised the Women's Reserve for its discipline and skill. The best tribute to Waves, he said, was the continuing requests for more of them.

As World War II drew to a close, members of the Women's Reserve joined the mass exodus from the military. Taking their newly learned skills with them, most Waves returned to civilian life in early 1946. A small nucleus remained as the service began its drive to make the Women's Reserve a permanent part of the peacetime Navy. In 1948, Congress passed the Women's Armed Forces Integration Act permitting women to join the regular Navy and the Naval Reserve.

By releasing desk-bound officers and enlisted men for duty afloat, patriotic Waves contributed to American success in World War II. Fortunate in their leadership, they efficiently handled the unusual as well as routine slots open to them and paved the way for women to be a permanent part of the U.S. Navy. With seriousness and dedication, Waves successfully fulfilled their roles as pioneers. Modern women can reflect with pride on the accomplishments of the trailblazers of World War II.

Dr. Susan H. Godson received a B.A. from George Mason University and M.A. and Ph.D. degrees from the American University. She is a historian and writes on 20th-century U.S. diplomatic and naval history. She and her husband William live in McLean, Virginia.

5

"The Navy Waves the Rules"

Ensign Mary E. Heckathorn, USNR

U.S. Naval Institute *Proceedings*
(August 1943): 1082–84

DID OLD COMMODORE JOE FYFFE live a hundred years before his time? It appears that he may have started something, when he wrote:

U.S.S. Monocacy, 3RD RATE,
WUSUNG, CHINA.
(1875)
Honl. Geo. M. Robeson,
Secretary of the Navy,

SIR:

I am given to understand that my fair country women in these seas, consider that I am wanting in the graces and accomplishments, necessary to a perfect character; that I am rough and bearish; and in fact "little better than the wicked."

I have too much respect for these ladies to doubt the justice of their sentence. I humble myself before them and promise by God's blessing and a little assistance from the gentle sex to become, in time, everything that they can wish.

In order to enable me to fulfill this vow, I request, in accordance with the admirable custom growing so much in fashion in the Navy, that my family may be permitted to live with me on board this ship.

Also my grandmother.

Not that I have the honor to possess such a venerable relative, but my wife has one she will lend me with pleasure; who will bring among us the stately and ceremonious manners of that indefinite period called "Old Times"; and who will prove a chastening corrective to all hands of us, for she has, for about a hundred years, cultivated that natural flow of eloquence so charming in her sex; and a soft answer never turns away her wrath, for she is as deaf as a post.

Very respectfully,

Your obedient servant,

JOSEPH FYFFE,

Commander, Comdg.

A feminine voice travels over the wires in answer to the ringing of the telephone, "Personnel Office, Miss Smith speaking."

"What did you say?" a bewildered masculine voice requests.

"Miss Smith, sir."

"Look, I want to speak to the officer in charge of the section."

"I'm she, sir."

Conversations like the above are heard every day in every office where the Women Accepted for Volunteer Emergency Service have begun their duties. Though they cannot be taken to sea as Commodore Fyffe suggested, the WAVES, as they are commonly called, are affecting the Navy, for they are not merely assisting officers and enlisted men, they are actually taking their places, doing the same tasks and receiving the same pay. The replacement is possible as the result of one to nine months of rigorous Navy training.

In many cases the women have learned their duties from the very men whom they are replacing, and have the satisfaction of being able to say, "I know the man whose position I am filling. I made it possible for him to go to sea." Certainly no one of the WAVES asks more than to be able to fill capably a Navy man's place on shore; the resulting satisfaction is worth every step of drill and every minute of study that she must do to complete her training.

The WAVES are typical, civic-minded American citizens who had been working before the war in any of the large number of fields open to women in the United States. They are not necessarily glamor girls (those who were soon fall into line once the training begins), but they are a healthy, wholesome, attractive group who possess, above all, the urge to serve their country. They also have enough of the pioneering spirit to try a new way of life.

Thus, taking little with them but patriotism and a pioneering spirit, the volunteers for the WAVES leave their former fields of work and undertake the adjustment to a military existence. The organization has attracted women from every field of endeavor. In a typical company at a training school, the fields represented ranged from teachers of all subjects from mathematics and science to music and art, from personnel workers, social workers, and secretaries to camp directors, chemists, librarians, home economists, airplane pilots, lawyers, and editors. Hence the women entering the United States Naval Reserve come from as varied backgrounds as the men of that branch; indeed, they are their sisters, daughters, and sweethearts. In many cases they are descendants of old Navy families.

Coming from so many fields of activity the prospective officers and enlisted women are rapidly fitted into Navy life. By the time indoctrination and the training period are finished the individualists are either weeded out or have adjusted their individuality to conform to Navy requirements. The thoroughness of this training can best be illustrated by the fact that not only is the care of the uniform taught but the posture of the wearer is checked and, where necessary, remedial measures are suggested. On the field the hand salute is practiced for hours to insure perfection. Marching is done with precision, and Navy customs are firmly implanted. To insure pride in the Navy code of conduct, time is spent in studying John Paul Jones's inspired "Qualifications for a Naval Officer," and "The Laws of the Navy" by Hapgood. In addition to military etiquette and appearance the trainees learn the Navy method of procedure in the specialized field they are going to enter. Thus, at the completion of the training the women fit easily into the routine of the station to which they are assigned.

The training program in itself is difficult, but the regular hours and well-planned schedules aid in maintaining enthusiasm; however, the high morale is maintained chiefly by esprit de corps. Pride in the platoon, the company, and the Navy develops in each and every woman in training. This spirit is inspired by the long marches on the drill field, the miles of trudging to class through sun, rain, and snow, the hours of study, the rigorous personal and room inspections, and the knowledge that as one training school song puts it, "This is the Navy." This esprit de corps, the by-product of the life in the training school, in the end becomes the force that has the most effect in changing the trainees from civilian to military personnel. It makes the uniform not simply raiment to be worn as a patriotic gesture for the duration, but a garb of distinction to be worn by a few fortunate women who have succeeded in living up to rigorous Navy standards. Whatever the future may hold in the line of duty, the WAVES find, before they complete their training, that they are ready and eager to do what the Navy expects of them. They have the proper training to do their work and, more important, they have gained the proper pride. They are Navy.

Mention should be made of the fine spirit shown by the instructors in the training stations for the WAVES. They, too, have been carefully selected for that most important duty of any officer, the training of junior officers. The instructors' attitude reflects consideration for and interest in making the training of the women thorough and adequate. Any deficiency in knowledge of Navy procedure should not reflect on the training given the WAVES; the teachers are intent upon making the women good Navy personnel.

In fact, that intent is so evident throughout the schedule in a training station that members of the WAVES soon criticize each other if any of their group fails to use Navy terminology or conduct herself as a good member of the Navy. Competition in conduct and studies became so keen at one station that the custom of posting grades was abolished.

Studies vary with the type of training given. Indoctrination includes Navy customs and traditions, naval history, organization, law, and ships and aircraft. The advanced courses include training in disbursing, supply, meteorology, communications, and further training in Navy organization for officers, while enlisted personnel are trained to be yeomen, storekeepers, radiomen, parachute riggers, Link trainer operators, and other types of technicians. The kinds of shore duty the women are entering appear to be limited only by the facilities available for training them. On at least one occasion no less an authority than Rear Admiral Jacobs, Chief of Naval Personnel, has said in effect that as soon as the WAVES have taken the places of the men and succeeded in one branch of the service they are requested for another. Branches of the Navy which first questioned the place of women in the organization have recently been requesting WAVES as shore duty replacements.

Having completed their training, the WAVES report to their duty station. The welcome they receive varies. In one case a WAVE was given a temporary assignment in an office where the commanding officer was opposed to women officers. At the completion of her duty he requested that a WAVE be assigned permanently to his department. That instance is typical, and the women realizing that they are on trial are anxious to do their work well.

In direct contrast, however, at many stations the WAVES are given a reception that they had in no way expected. Officers and enlisted men, eager to go to sea, do everything possible to help their feminine replacements learn their tasks on shore and to make them happy in their work.

Whatever the initial attitude is at any station the WAVES have, in a short time, found the men accepting them as fellow members of the Navy. In fact, some offices have even sent WAVES on field trips to acquaint them with the work in outlying branches.

Those trips have actually included outposts where no women had ever been allowed to enter. Even there the WAVES were welcomed as officers in the Navy and shown the activities of the station; any questions they asked were answered.

A trip which will remain in the memory of the ten WAVES who made it, was one aboard a Navy ship for luncheon. Having entered an area where only Army and Navy personnel, and a few especially privileged civilians can obtain entrance, the group was driven to the starboard gangway. Boarding the ship took place according to Navy custom, to the accompaniment of the shrill whistle of the boatswain's pipe. To the evident satisfaction of the men officers the WAVES showed their knowledge of the proper boarding etiquette. The courtesy of the officers at luncheon and during the tour of the ship gave evidence that the WAVES were women guests—yet, when the women asked questions it was equally evident that the men accepted them as fellow officers, freely discussing military matters with them. Women, the WAVES were, but officers in the Navy also, and the thrill of belonging to the latter group was evident in the pride of bearing of each WAVE as she went ashore and paused to salute the Officer of the Deck and the colors.

In these times when every woman is exhorted to perform to her utmost any task of which she is capable, in the furtherance of the war effort, it is felt that every WAVE understands how Admiral A. P. Niblack, U.S. Navy, felt when he wrote his son, "Personally, I would rather be a commissioned officer in the Navy than hold any other position under the Government. It is an honor; it is a career; it is one of the most exacting and difficult of all professions. Otherwise, I would not have urged you to enter the Navy."

"WAVE Training"

6

Captain J. L. Woodruff, USNR

U.S. Naval Institute *Proceedings*
(February 1945): 151–55

IN THE LIGHT of recent public announcements of the large numbers of WAVEs now on duty in the shore establishment of the Navy, and because of the praise of their accomplishments by high ranking officers, it is felt that service personnel may be interested in a brief account of the training and indoctrination of WAVE officers.

The U.S. Naval Reserve Midshipmen's School (WR), located at Northampton, Massachusetts, is, as its title indicates, a training school for officers of the Women's Reserve. It employs some of the facilities of Smith College, as well as the Hotel Northampton (dormitory) and the famous Wiggin's Tavern for messing. And what messing! Nowhere in the country, it is believed, do any trainees receive, within the limits of the standard Navy ration, such delicious meals, of such interesting variety, and cooked to perfection as they would come out of the most skilled housewife's kitchen.

NRMS (WR), as we shall call it for brevity, is, in fact, a miniature Naval Academy for women. Insofar as the length of the course and other factors permit, it is organized and administered precisely as is the Naval Academy. Under the Commanding Officer, the Officer in Charge of Midshipmen, a Lieutenant Commander of the Women's Reserve, performs the functions of Commandant of Midshipmen. Through the Regimental Commander, a WAVE lieutenant, she supervises the receiving, billeting, quartering, messing, uniforming, drill, and discipline of seamen and midshipmen.

The Regiment consists of two battalions of two companies each. Each company contains four platoons, two each of seamen and of midshipmen. The number in a platoon is from 25 to 35. Commissioned WAVE officers are detailed to duties as Battalion and Company Commanders, assistants, and as Regimental and Battalion Aides. The duties of these officers are administrative and supervisory, corresponding to similar duties of officers of the Executive Department, U.S. Naval Academy.

As at the Naval Academy, midshipmen officers, selected and appointed on the basis of aptitude as well as scholarship, carry out all possible details in connection with shakedown, training, drill, and discipline of the regiment. They assist in organizing and carrying out the intricate routine of receiving new seamen, getting them uniformed, and installing them in quarters. They teach them to stow lockers and make up bunks properly, to care for their rooms, and all of the little routines of military life, so simple to the initiated, yet so baffling to the recruit.

Comparable to the Academic Departments of the Naval Academy is the Instruction Department, which supervises the curriculum of academic subjects. Of these there are five, viz., Naval Organization, Personnel, History and Strategy (including Current Events) and Law, Ships and Aircraft, and Communications and Correspondence. This department is staffed entirely by Women's Reserve Officers, especially selected for their teaching ability, intelligence, and personality. At present about thirty instructors are employed. A policy of rotation with graduates who have gained actual experience in field billets of the Women's Reserve is followed, in order that the content of courses and teaching may be authentic and up to date. For the same reasons, fifteen-day field trips for instructors are ordered, during which they inspect activities and gather material appropriate to their respective subjects. The inclusion of fresh information thus gathered in the several courses keeps them from becoming stereotyped and obsolescent, as well as preventing instructors from becoming stale and ineffective.

Nor is the physical side of training overlooked. The Physical Education Department, under a trained and experienced WAVE officer, supervises about five hours each week of physical training suitable for women, including exercises calculated to develop good posture, trim figures, grace, and stamina. The improvement in these characteristics observed during the brief training period of seven weeks is notable to the point of incredibility. This program includes sports such as soft ball, volley ball, badminton, archery, and hockey, with occasional Field Days for competition and fun.

Approximately 260 women enter NRMS (WR) every four weeks for training. About a fourth of these are enlisted women of various ratings, recommended for this training by their Commanding Officers as a result of their excellent performance, and having the necessary backgrounds of education and experience to warrant officer training. The remaining three-fourths is composed mostly of college graduates, the majority in their early twenties. A fair number of older women, mostly college graduates with considerable administrative or technical experience, is included. These are in training for commissions in the grades of lieutenant, junior grade, and lieutenant, senior grade, many for specialized billets for which they were procured.

In fact, for officers who are to perform certain special duties, such as supply and disbursing or communications, indoctrination at NRMS (WR) is only the first stage of training. On graduation and commissioning, officers previously selected for such duties are sent to the Navy Supply Corps School (Radcliffe Branch) at Harvard University, Cambridge, Massachusetts, or to the Communications School (AV). The latter school is a separate activity at NRMS (WR), under the same command, and utilizing the same facilities. Graduates of NRMS (WR) have also been sent to other specialized training on occasion.

Since the entire course of training covers a period of eight weeks, this results in the graduation and commissioning of some 250 officers of the Women's Reserve every four weeks. While the attrition is relatively small, this is not due to low standards of achievement. On the contrary, standards are exceptionally high and are rigidly enforced. The Academic Board meets twice on each class. Before it come the name and record of every student who, by reason of low scholastic grades, questionable aptitude, or other observable characteristics, may, in the opinion of the board, be deemed to lack the full requirements for an officer of the Navy. The reasons, however, why attrition is low are simple. First, the motivation of those who enter is exceedingly high. There is no problem of forcing or inducing women to put forth their best efforts. The problem, rather, is to control over-conscientiousness so that it does not become a mental hazard. Then, too, many of the entrants are recent college graduates, carefully screened. Their intelligence is high, and their assimilation of material presented, after they accustom themselves to a strange environment and different methods of instruction, is rapid. Finally (a distinctly feminine trait) their ideals of service are so high that nothing can swerve them from their goal.

As is the case at the Naval Academy, the stress in training is laid on the development of that intangible thing called military character. Every opportunity is taken to bring out or inculcate qualities of initiative, moral courage, resourcefulness, integrity, and self-reliance. At first thought, it might be felt that the brevity of the course does not permit as much of such development as might be desired. The salvation, of course, is in the quality of the trainees. To the maximum extent possible, however, responsibilities of leadership and performance are thrust upon midshipmen as rapidly as they are qualified to assume them. They are required to drill their battalions, companies, and platoons; march them to and from classes and meals; train their juniors as previously mentioned; stand watches and perform special duties. During the graduation week, a substantial number of the class not graduating is assigned to various departments and activities of the station proper, where the midshipmen understudy various officers in their duties and gain some comprehension of the organization and administrative problems of the school. Others assist in the "shakedown" of entering seamen, relieving battalion and company officers of a great portion of the arduous detail involved.

In all these duties, and in practically every activity of their daily lives, midshipmen and seamen are carefully observed for the possession and development (or want of) those qualities of leadership and military character considered essential for officers.

Routine and special reports on aptitude are submitted by all regimental, instruction, and departmental officers on students coming under their observation. These reports are studied and evaluated; interviews are held to clarify doubtful points and appraise personality traits; and, in general, a searching study is made to evaluate the worth of each individual to the naval service. Finally, no midshipman is permitted to graduate and be commissioned who is not deemed worthy to take her place as an officer—who cannot direct herself as well as others, and work in harmony with both male and female commissioned and enlisted personnel of the navy, within the Navy pattern.

The period actually available for training and instruction is slightly less than seven weeks. In this time must be compressed drills, lectures, physical education, uniforming, classes, study, classification interviews, physical examinations, inoculations, recreation, and even a little liberty, as well as the daily requirements of just living. It is a crowded and fast moving schedule. Obviously it would be impossible to treat in great detail the five academic subjects previously mentioned. In the approximately 200 hours available for study and recitation it is possible only to cover the highlights; to give each student a broad generalization of the Navy's requirements; and to provide her with a knowledge of the systematic, Navy method of approach to the multifarious problems she will encounter as an officer. She is given a complete, over-all picture of the organization of the sea and shore establishment of the Navy, as a means of understanding the interlocking authorities and responsibilities involved in official acts and procedures. She is given a thorough grounding in personnel matters, that she may recognize and know ranks and ratings, functions and duties; that she may understand the basic principles involved in handling subordinates, as well as in dealing with equals and seniors. She is introduced to a fast moving account of naval history, and acquainted with the traditions and customs of our Navy. She is made conversant with the backgrounds of the present world struggle and is brought up to date on its progress and implications. She is acquainted with the fundamentals of ships, aircraft, and ordnance, in order that she may deal more intelligently with the materiel of the Navy, as well as speak its language. She is introduced to the rudiments of naval law, discipline, and punishments, not as a threat, but to aid her in guiding others as well as to perform her military duties intelligently. She learns to compose and to comprehend the meaning of dispatches and other communications, to be of more value to her seniors; and she learns to prepare, route, read, and act on Navy correspondence, that she may implement her own ideas, as well as those of others.

If, as a result of these concise and streamlined courses, the new WAVE officer is able to quickly adapt herself to the requirements of the billet in which she finds herself, to converse and act intelligently on naval matters within her scope, and to deal with naval personnel in a military yet human way, her training has accomplished its purpose.

A short chronological account of the progress of a typical class may be interesting. Arriving by train or bus on Thursday, Friday and Saturday, entering candidates are processed through the assembly line (orders, forms, etc.), formed into small groups, and taken to their quarters. Here they are settled, given preliminary instructions, and introduced to the requirements of military life. They are taken to the uniform shop to be fitted for uniforms, given sufficient military drill to be able to march acceptably in formation, and generally "shaken down" and settled in their new environment. In this connection, a new term, used (we trust, inadvertently) for the first time by one of the staff on a radio broadcast, has been invented, and is deemed worthy of the Navy's consideration. The term is "squartered," meaning "settled in quarters."

During this period of "shakedown," the class already aboard assists. No academic activities are scheduled, except certain conferences on a wide variety of topics for students not otherwise engaged.

All candidates enter the school as apprentice seamen. Enlisted women (Class V-10) entering become apprentice seamen, regardless of their previous rates, until their status as midshipmen is determined.

The following three weeks are devoted to that portion of the indoctrination course which may be called basic, although the courses now given are continuous and undivided. The academic subjects previously listed are studied and recited on; infantry drills and physical education are conducted; reviews are held on Saturdays, alternating with field days; and the seamen become accustomed to military life.

Because of distances involved between quarters and classrooms it is not feasible to have alternate study and recitation periods as at the Naval Academy. Hence the instruction schedule is quite intricate, and the task of integrating physical examinations, inoculations, fitting of uniforms, ID photographs, and a myriad other activities becomes an almost Herculean labor. Coupled with this, messing facilities and shortage of help make it necessary that each battalion mess at a different hour. The problem thus becomes one of operating two schools, each on a different time schedule, yet avoiding overlapping or wasted time. This takes ingenuity, since the most insignificant (apparently) change has widespread and sometimes almost disastrous results affecting every other activity.

During the first three days of the fourth week (which is graduation week for the class completing its course), the seamen take their first examinations. The marks for these, with the daily grades accumulated, and an aptitude report on each, form the basis on which the Academic Board will consider apprentice seamen for discharge, return to V-10, or further retention. Those found wanting are discharged, if entered from civil life, otherwise returned to enlisted status (V-10) in their former ratings; those considered and retained are frequently placed under observation, especially if reviewed for questionable aptitude.

During the latter half of this graduation week these seamen about to become midshipmen assist with the "shakedown" of the new class, understudy departmental and staff officers, and attend the conferences previously mentioned. At this time also are chosen the new midshipman officers who will perform regimental duties during the ensuing month. At the beginning of the following week all of these seamen are sworn in as midshipmen, and put on their midshipman hats. This is the greatest event of their course to date.

The next three weeks are somewhat a repetition of the first three, except that now the midshipmen are going down hill. Their studies are more advanced; their duties are more like those of officers; and they are under stricter observation, if possible, for qualities of leadership and military character. They are being fitted for more uniforms, and generally preparing for the fateful day of graduation. In this last week come examinations again, and the Academic Board meets to make final determination as to who, if any, must fall by the wayside. Now the orders begin to come in, and the prospective graduates are agog to learn their future billets, and busy preparing to "shove off" and take their places as officers of the Navy.

And then, graduation day dawns! A review which compares favorably in quality, if not in numbers, with the Naval Academy dress parade, begins the ceremonies. Then come the graduation exercises, with a salutatorian address by a midshipman chosen for scholarship and ability; the graduation address by a visiting dignitary; and last, but not least, the oath of office by which the graduating midshipmen become officers. This is the climax, and a fitting one. It remains only for the new officers, eyes turned toward the new stripes, to get under way for the stations mentioned in their orders, while those they leave behind them become almost as enthusiastic in the realization that they now have only "one more river to cross."

Captain Woodruff is a graduate of the Naval Academy, Class of 1917, and served afloat and ashore until his resignation in 1924, when he immediately entered the Naval Reserve. An authority on Marine and Air Navigation, he was an instructor at the Naval Academy for two years until detailed to the Navigation Training Section of the Bureau of Aeronautics in 1942. At present he is District Operations Officer and Port Director on the staff of the Commandant of the Ninth Naval District.

"Women in the Sea
Services: 1972–1982"

7

Captain Georgia Clark Sadler, USN

U.S. Naval Institute *Proceedings*
(May 1983): 140–55

*It wasn't so long ago when the idea of a woman quartermaster
doing navigational plotting in a seagoing ship would have been
unthinkable in "this man's Navy." But much that was previously unthinkable has now become fact. It is no longer just "this
man's Navy"—or Marine Corps or Coast Guard, for that matter. Many—though certainly not all—of the old barriers have
been falling. The landmark changes of the past decade have
made women an integral part of the sea services and contributed substantially to overall readiness. Women still do not have
a combat role, but they now do serve in ships, hold command,
fly airplanes, attend service academies, and work alongside men
in any number of nontraditional billets. At a time when available
manpower is shrinking, women are being accepted for the most
practical of reasons—they are needed.*

THE TEN YEARS from 1972 to 1982 marked a dynamic decade for
women in the Navy, Marine Corps, and Coast Guard. In 1972, their
function was to provide a flexible, well-trained cadre which would permit rapid and efficient mobilization of large numbers of women in event

of another war. By 1982, women had become a significant, integral part of the sea services without whom the operational readiness of the forces would suffer. Between 1972 and 1982, the number of women officers almost doubled, while the ranks of enlisted women increased nearly five-fold. New fields opened to women as the services significantly expanded their utilization. Every year, new historical milestones were passed.

With the women's liberation movement gaining strength and the Equal Rights Amendment having passed Congress in 1972, the military, along with the rest of American society, was under increasing pressure to expand opportunities for women. Additionally, the drive for change was reinforced and accelerated by legal challenges to laws and military policies concerning women.[1] The paramount reason for the changes, however, was the manpower requirements of the services. In the case of the Navy and Marine Corps, the specific push came from the introduction of the all-volunteer force (AVF) on 1 July 1973. Military planners were told to increase the use of women to offset any shortage of men resulting from the end of the draft. Also, as the decade continued, it became apparent that the number of men in the manpower pool from which the Navy and Marine Corps would draw would decline during the 1980s and would not increase again until the 1990s; therefore, they began to look to all personnel resources from which they could recruit, not just "manpower." Though not part of DoD, the Coast Guard faced much the same recruiting outlook.

In addition to the momentum generated by equal opportunity, court cases, and manpower requirements, specific impetus was provided to the Navy and Marine Corps by the Office of the Secretary of Defense (OSD) during all administrations between 1972 and 1982. In April 1972, the Nixon Administration directed the services to ensure that women were given equal opportunity and to eliminate treatment and regulations which precluded adequate career opportunities for women. The Carter Administration was especially aggressive in setting numerical goals for each service, and it was between 1977 and 1981 that much of the major growth in the number of women occurred. The Reagan Administration also recognized that women are a key to the success of the AVF and not returning to the draft. In early 1982, Secretary of Defense Caspar Weinberger told the service secretaries that the administration wanted to increase the role of women in the military and to break down any institutional barriers preventing the fullest use of women.

Characteristics

Anyone who visited a Navy, Marine Corps, or Coast Guard base in 1972 would have been struck during a return visit in 1982 by the increase in the number of women in uniform working at the various activities. By far, the largest numerical increase was in the Navy. In 1972, slightly under 6,000 enlisted women were on active duty, and of 100 sailors, one was a woman.[2] By 1982, the number had increased to 37,000, and eight of every 100 sailors were female. The number of Navy women officers rose from slightly over 3,000 in 1972 to about 5,750 by 1982; the proportion of women in the officer corps more than doubled, from 4% to 8.5%. In the Marine Corps, the number of enlisted women more than tripled from almost 2,100 in 1972 to about 7,900 in 1982. Of 100 Marines in 1972, one was female; in 1982, four were women. Among Marine Corps officers, the number of women rose from about 260 in 1972 to 560 by 1982, expanding from 1% of all officers to 3%. In relative terms, the largest increase took place in the Coast Guard. It entered 1972 with just two women officers and 25 enlisted women, and all were reservists on extended active duty. By 1982, the Coast Guard had about 130 women officers and almost 1,700 enlisted women on active duty, and women comprised 5% of the enlisted strength and about 3% of the officer corps.

The sea services were able to increase their numbers because they were very successful in recruiting the quantity and quality of women they wanted. Of 100 individuals who enlisted in the Navy without prior service, the number who were women rose from two in 1972 to 11 in

1982. In the Marine Corps, the number tripled from two out of 100 in 1972 to six in every 100 in 1982. In terms of quality, because Navy and Marine Corps enlistment standards were higher for women than men, the proportion of enlisted women who have high school diplomas and were in the top three categories of the Armed Forces Qualification Test (AFQT) was consistently higher than for men.

Among Navy and Marine Corps women officers, the major change was the source of their commissions. In 1972, most Navy women received direct commissions or attended Officer Candidate School (OCS). In 1982, these still were the major sources of women officers, but women were also receiving commissions through Aviation OCS, the Naval Reserve Officer Training Corps (NROTC), the U.S. Naval Academy, and the Limited Duty Officer (LDO) program. Similarly, the Marine Corps' accession sources changed with the addition of NROTC and Naval Academy graduates.

The Coast Guard differed from the other sea services in its recruitment of women. In 1972, the Coast Guard was not recruiting women for active duty service. In 1973, enlistment of women for four-year tours of active duty was authorized, and women began attending the Coast Guard's OCS. Initially, similar to the other sea services, the Coast Guard limited the number of women who could be recruited and the occupations they could enter. In 1978 however, the Coast Guard abolished setting annual numerical ceilings for the recruitment of women and opened all skills to them. Also, the Coast Guard's enlisted recruiting standards were the same for both men and women. Finally, in 1980, the Coast Guard Academy was added as a commissioning source for women officers.

The relatively large numbers of the women who came into the sea services during the 1970s were still in the lower and middle paygrades by 1982. In 1982, only 7% of enlisted women were in paygrades E-6 to E-9, compared with 16% of male Marines and 25% of Navy men. Women were a junior force, because they generally did not have sufficient time-in-service to meet promotion requirements to the senior grades. For example, the average male sailor had 50% more years in the Navy than the average woman. Promotion data, however, indicated that eventually the women's paygrade structure should approximate that of the men.

Women officers were also found primarily in the junior ranks, especially in the Coast Guard where the senior woman in 1982 was a lieutenant commander. One major change did occur during the decade—Navy and Marine Corps women were appointed to flag and general officer rank. The Navy generally had two women rear admirals, one nurse and one line officer. The Marines appointed one woman brigadier general, who served from 1978 to 1980.

While the number of women coming into the sea services was increasing, the occupational fields open to them also expanded. In 1972, enlisted women could enter only about one-fourth of the Navy's and Coast Guard's ratings (occupations) and somewhat over one-half of the military occupational specialties (MOS) in the Marine Corps. By 1982, those figures had risen to 86% for the Navy, 96% for the Marine Corps, and 100% for the Coast Guard. Certain Navy and Marine Corps skills have remained closed, however, because of their combat relationship or in order to maintain acceptable rotation patterns for men. With the additional job opportunities, the proportion of women in nontraditional skills increased from zero in the Coast Guard and a little over one-tenth in the Navy and Marine Corps in 1972 to somewhat under one-quarter in each of the services by 1982. Even with the growth, about half of all enlisted women were in traditional skills; the remainder had not been designated.

Among women officers, major changes in occupations also occurred. In 1972, about three-quarters of women Navy officers were nurses, with most of the remaining one-quarter in the unrestricted line. During the decade, the number of women in the general unrestricted line[3] grew, the restricted line and remaining staff corps were opened to women as were all of the warfare communities, except submarine and special warfare which remain closed because of legal restrictions. By 1982, the proportion of women officers who were nurses had dropped to less than half, while the percentage in the unrestricted line had risen to 45%.

Because the Navy provides the Marine Corps with its Medical support, no women Marines are nurses. Nevertheless women officers in the Marine Corps continue to be found primarily in traditional occupations. In 1981, of the women officers with designations, more than one-half were in the administrative occupations and somewhat under one-quarter in supply and logistics. The remaining women were about evenly divided among the other occupations, except combat arms and pilots which are closed to women.

Along with the opportunity to enter new skills, women were also assigned to new types of commands and locations. In 1972, enlisted women served primarily at shore-based activities in the continental United States with a small proportion overseas. In the following years, women were ordered to a number of new types of commands, including operational aviation and seagoing units and the Fleet Marine Forces (FMF). In terms of locations, women were assignable worldwide, and by 1982 virtually all geographic locations were open to them. For example, in 1982 the Navy began permanently assigning women to commands located on isolated Diego Garcia in the Indian Ocean. By 1982, about 30% of Navy enlisted women were overseas and 8% were on shipboard duty. Between 1977 and 1982, the number of women Marines serving in the FMF increased three and one-half times.

Even with the ever-increasing numbers, the proportion of Navy and Marine Corps enlisted women who did not complete the first three years of their enlistments generally decreased.[4] A major reason for the decline was that beginning in 1975 women who became pregnant were no longer involuntarily discharged. In comparison with men, while the proportion of Navy women not finishing the first three years of service was roughly the same, the percentage for women Marines was considerably higher. Concerned about the loss of these women, the Marine Corps had a study under way in 1982 to determine the reasons for the high attrition rates. Among women completing their first enlistments, the proportion reenlisting was higher than for men until 1977; thereafter, women's first-term reenlistment rates were about the same or lower than those of men.[5] The sharper decline in the women's rates can be attributed, at least partially, to the changes that were occurring. For example Navy women who joined when they comprised a small elite portion of the force stationed almost exclusively in the continental United States began to find themselves part of a much larger group facing overseas duty at isolated locations and, after 1978, shipboard duty. Consequently, some decided not to reenlist.[6]

Finally, between 1972 and 1982, major shifts occurred in the marital and parental status of women. In 1972, Navy women were overwhelmingly single, and almost none were parents. Generally, women who married left the Navy, and those who became pregnant or adopted children were, almost without exception, involuntarily discharged. By 1980, although most enlisted women were still single, the proportion who were married had jumped to 45%. Of the married women, almost two-thirds of the officers and about half of the enlisted women had civilian husbands. Of women with husbands in the military, 9 out of 10 were married to someone in the Navy and the most common pattern was marriages between enlisted Navy women and enlisted Navy men. Officer/enlisted marriages were comparatively rare. In terms of parenthood, 17% of Navy women had children.

Changes in Policy and Law

The policy message on women issued by the Chief of Naval Operations, Admiral Elmo R. Zumwalt, Jr., in August 1972 was a watershed for women in the Navy. In that message, Z-gram 116, Admiral Zumwalt indicated that while women in the Navy had historically played a significant role, he believed the Navy could do far more than it had to accord women equal opportunity. He also recognized the heightened importance of women as a vital personnel resource in the all-volunteer force. As a direct result of this Z-gram, many new fields and opportunities opened to women. Limited entry of enlisted women into all ratings was authorized, and a pilot program was initiated for the assignment of women to the crew of the hospital ship *Sanctuary* (AH-17). Women began to be assigned as commanding officers and executive officers at shore activities. All the staff corps and restricted line communities were opened to women, as was the NROTC program. The pattern of assigning women exclusively to certain billets was eliminated, and qualified women were to have tours covering the full spectrum of challenging and highly career-enhancing billets, including briefers, aides, executive assistants, service college faculty members, action officers on the CNO's staff and the Joint Staff, senior enlisted advisors, etc. Women were also to be assigned to duty at more operational types of commands, for example, major fleet and type commanders, operational training commands, and antisubmarine warfare tracking facilities. Joint service colleges were to have Navy women as students and faculty members. As a personal illustration of these changes, almost every assignment the author has had since 1972—instructor at the Naval Academy, intelligence briefer to the Chairman of the Joint Chiefs of Staff, action officer on the Joint Staff, and student at the National War College—would have been impossible prior to Z-gram 116.

The major policy change for women in the Marine Corps occurred in 1974 with their assignment to units in the Marines Forces (FMF). In the early 1970s, the Marine Corps, like the other services, was having difficulty recruiting the quantity and quality of men it required. The shortages were especially critical in the technical occupations, many of which were in the FMF. Meanwhile, the number of women was increasing, and some of those women were entering the technical skills. The Marine Corps, therefore, decided that at least part of the shortages in the FMF could be made up by employing women in units not routinely exposed to combat action. A pilot study in 1974 called for the eventual assignment of about 70 officers and 450 enlisted women to the FMF; by the end of this year, 13 women were assigned to nine units. A follow-on review in 1977 indicated that the FMF could utilize up to about 3,700 women. By 1982, the number of women assigned to combat support and combat service support units in the FMF had grown to almost 600 women in 39 units.

For the Coast Guard, the major policy changes came in 1977 and 1978. By virtue of being in the Department of Transportation, the Coast Guard is not bound by the same legal restrictions as the Navy with regard to assigning women to ships. Consequently, in May 1977, the Coast Guard began assigning women to sea duty, and by 1982, 124 enlisted and 35 officers were assigned to units afloat. In 1978, all restrictions were removed on assignments, specialties, training, and command opportunity for women. In 1982, for example, two women were commanding officers of patrol boats and one was executive officer of a buoy tender. Although the Coast Guard makes every effort to assign women to units in groups of two or more for medical and companionship reasons, it does not deny an assignment solely because of a lack of a second woman. Physical limitations with regard to privacy in berthing and personal hygiene do result, however, in women not being ordered to certain geographic areas and to some seagoing vessels.

Meanwhile, substantial advances occurred in the aviation field. In 1973, the first women entered the naval flight training program. By 1982, 90 women were pilots or pilot trainees, and they were flying most of the aircraft in the Navy's inventory. A number of women pilots became carrier qualified and performed carrier-on-board deliveries. In 1982, the first woman was selected for test pilot school. In 1979 the naval flight officer (NFO) program was opened to women, and 31 women were NFOs or NFO trainees by 1982. Meantime, women officers not in the aviation communities were also assigned to squadrons in non-flying billets, such as intelligence officer. In 1979, a test program was begun by assigning 45 enlisted women to non-flying billets in four antisubmarine patrol (VP) squadrons. Finally, women assigned to fleet replacement squadrons began to go on temporary duty aboard carriers as part of the squadrons' carrier qualification detachments.

The Coast Guard also opened its flight program to women and in 1982 had three women pilots. The Marine Corps, on the other hand, continued to restrict women from assignment as pilots or crew members.

The restriction is based on the law prohibiting assignment of women to aircraft with a combat mission and the Marine Corps's belief that assignment of women to a combat role is inappropriate. The vast majority of its aircraft are combat aircraft, with very limited numbers of support aircraft. The Marine Corps maintains that assignment to the small number of support aircraft would be a mistake for two reasons: first, male pilots would be deprived of rotation through these assignments and, second, women cannot be rotated in these capacities to any other assignments.

The decade also saw the demise of the separate management of Navy and Marine Corps women. In 1972, a quasi-chain of command existed for Navy women consisting of women representatives at the command level, assistants for women at naval districts and, in Washington, the Assistant Chief of Naval Personnel for Women (Pers K), also known as the Director of the WAVES. In 1972, a pilot program was announced to suspend assignment of women officers as women's representatives and assistants for women; the assignments never resumed. Additionally, a major effort was undertaken to discourage the use of the acronym WAVES (coined in World War II to stand for Women Accepted for Voluntary Emergency Service) to refer to women in the Navy since the name did not accurately reflect the current concept of women as full, permanent members of the Navy team. In 1973, Pers K was disestablished. In the Marine Corps, the office of the Director of Women Marines was disbanded in 1977.

Another major area of change was the pregnancy separation policy. At the beginning of 1972, the policy in effect stipulated that if a woman became pregnant or adopted a child, she was not to be allowed to remain in the Navy or Marine Corps.[7] In February of that year, the policy was modified to permit women to request waivers to stay in the service. Generally these waivers were granted. In 1975, as a result of a change in the Department of Defense (DoD) policy on pregnancy separation, the policy was reversed: instead of being required to get out unless they asked to stay in, pregnant women were to stay in unless they asked to get out. Normally discharge requests were granted on a routine basis.

As the years passed, circumstances changed, and the Navy and Marine Corps began to consider involuntary retention of pregnant women. Women were increasingly entering occupations or skills which required substantial training or education. Also, major growth occurred in the number of women in skills with manning shortages which made them eligible for large bonuses. Despite the investment, in most cases various laws and DoD policies effectively precluded the Navy from recouping monies paid for education, training, or bonuses if a woman left the service due to pregnancy prior to the end of her enlistment. Also, men increasingly voiced the view that the separation policy was inequitable and discriminatory, that is, it gave the women an "easy out." Consequently, in 1982, the policy was changed to enable the Navy and Marine Corps to retain some pregnant women involuntarily. Under the new policy, women normally are separated for pregnancy upon their request unless retention is determined to be "in the best interests of the service." Such retention is called for if women are in occupations or skills which have significant personnel shortages, have not completed obligated service incurred for funded education or training, have received a bonus, or have executed orders or entered a program requiring obligated service while pregnant. The policy also provides, however, that a request for separation can be approved for a woman in one of the above categories if she "demonstrates overriding and compelling factors of personal need" which justify her separation for pregnancy. Thus, over the decade, the Navy and Marine Corps shifted from involuntary separation to, in certain cases, involuntary retention of pregnant women.

Lastly, another major change was the Navy's elimination of separate enlisted recruit training and initial officer training for women. In the Marine Corps, although male and female officers began to train together, it did not integrate recruit training. The Marine Corps maintains that the scope of male recruit training includes subjects required to produce a basic Marine rifleman who is able to sustain himself on the battlefield, an objective not appropriate for women because of the combat restrictions. It also asserts that the physical standards of male recruit training are necessarily rigorous and that it cannot adjust, or appear to adjust, this training to adapt to women's capabilities, because that would undercut the basic training concept.

In addition to the initiation of new policies, five major legal changes occurred which affected women. While the services generally favored the changes, the Department of the Navy vigorously opposed one of them—admitting women to the service academies. During congressional testimony in 1974, Navy officials argued that the Naval Academy's purpose was to educate and train officers for combat roles, and since women were precluded by law from serving in ships or aircraft with a combat mission, they should not be permitted to enter the academy. Secretary of the Navy J. William Middendorf II testified that: "Simply stated unless the American people reverse their position on women in combat roles, it would be economically unwise and not in the national interest to utilize the expensive education and facilities of the Naval Academy to develop women officers. Therefore, it is imperative that education and training at that institution be reserved for those with the potential for combat leadership."[8] Those favoring admitting women disputed the services' claim that the academies trained officers exclusively for combat and cited the inequity of not providing full opportunities for women. In the summer of 1975, Congress voted to change the law, and in July 1976 the first women entered the Naval Academy. Meanwhile, in August 1975, the Coast Guard announced it would also accept women at its academy beginning with the class entering in 1976.

Unlike the Naval Academy legislation, the Navy and Marine Corps supported changing the legal restrictions on the assignment of women to ships. The law governing the assignment of Navy and Marine Corps women—Section 6015, Title 10, U.S. Code—prohibited duty on any ship, except hospital ships and transports. In 1972 the Navy did not have any transport ships, and the only hospital ship was the *Sanctuary*, which had been reactivated for the Vietnam War. After the war, the ship was decommissioned, and therefore the law effectively precluded women from sea duty of any sort, even on a temporary basis. Ironically, given the way the law was written, civilian women and women from the other services could be assigned to ships, but not Navy or Marine Corps women. Because the Department of the Navy did not want to raise the controversial question of women in combat, it recommended, as a first step, that the law be amended rather than repealed. It proposed to permit women to go aboard any Navy ship, including combatants, for temporary duty and to be permanently assigned to transports, hospital ships and "vessels of a similar classification." In their testimony, Navy officials indicated that a change in the law would provide expanded opportunities for women, help in Navy manning requirements and permit women Naval Academy midshipmen to receive training similar to that of the male midshipmen. The amendment passed Congress and was signed into law in October 1978.

Because the Navy planned for the assignment of women to ships for several years prior to passage, it was ready to implement the law immediately. Thus, one month later, in November 1978, the first five women officers reported aboard the repair ship *Vulcan* (AR-5). In December, enlisted women also began to report to ships. By 1982, 176 women officers were on board 31 ships, and almost 2,200 enlisted women were serving in 20 ships. Women were assigned primarily to tenders and repair ships, but they also were serving in one-of-a-kind ships, such as the training carrier *Lexington* (AVT-16), the missile test ship *Norton Sound* (AVM-1), and the deep submergence support ship *Point Loma* (AGDS-2). Alterations to the ships were minimal and usually involved ensuring women's privacy in the berthing areas and modifying the sanitation facilities. Another result of amending the law was that women officers were permitted to enter the surface warfare and special operations communities; 137 and 10 women were in each community respectively by 1982.

The third major legal change which affected women in the Navy and Marine Corps was the passage in late 1980 of the Defense Officer Personnel Management Act (DOPMA). DOPMA was designed to equalize the treatment of male and female commissioned officers by repealing all sections of the law which required separate appointment, promotion, accountability, separation, and retirement of women officers. It did not, however, repeal the combat exclusion provisions of Section 6015. The most significant impact of the law was in the area of promotions. Prior to DOPMA, with a few exceptions, women officers competed only against other women for promotion; under DOPMA they began to compete against men. Since the legal constraints on their assignments made women's career paths substantially different than men's, women expressed great concern that their promotion rates would drop sharply under the new system. Consequently, the Secretary of the Navy and the services undertook several actions to ensure equitable treatment of women before promotion boards, such as instructing boards not to allow women's different career paths to prejudice their selection and by monitoring board results very closely.

The major changes in the law for women in the Coast Guard resulted in their integration into the regular force. In 1942, the Women's Reserve was established as a branch of the Coast Guard Reserve. In 1947, the law establishing the Women's Reserve was repealed and all women were separated from the service. In 1949, the Women's Reserve was reestablished, but the number of women on active duty remained extremely small. In 1972, the Coast Guard convened a board to study the utilization of women. The board recommended expansion of the Reserve program with a view toward opening the way for women to enter the regular Coast Guard. In July 1973, the Coast Guard proposed legislation to abolish the Women's Reserve. In presenting the bill to Congress, it maintained that the Coast Guard no longer required a separate and distinct organization for women which operated apart from the mainstream of the Coast Guard. Rather, it argued, the times and needs of the service clearly called for the full integration of women; women should train and serve on both active and inactive duty as an integral part of the Coast Guard. Congress passed the legislation disestablishing the Women's Reserve in December

1973, and the first women were enlisted in the regular Coast Guard in January 1974.

While these four changes to the laws concerning women were brought about through the passage of legislation, the fifth resulted from a ruling by the Supreme Court. In 1972, the laws governing the dependency status of civilian husbands of women on active duty specified that a man could not be a dependent of a female service member unless he was dependent on her for more than half of his support. Furthermore, to qualify as a dependent, he had to be incapable of self-support because of a mental or physical incapacity. Even though, for example, a woman might be providing more than half of her husband's support because he was in college and not working, he still did not qualify as a dependent. The practical impact was that civilian husbands of active duty military women were not eligible for various dependent benefits, and the housing allowance (BAQ) received by married military women was at the single rate. In 1973, however, as a result of a suit filed by a woman Air Force officer, the Supreme Court struck down the law.[9] Thereafter, married Navy women began to receive BAQ at the "with dependent" rate, and their husbands were entitled to a spouse's identification card and the privileges accompanying it.

Combat Restrictions

The amendment of Section 6015 in 1978 permits the permanent assignment of women to ships and aircraft not having a combat mission and temporary assignment to any ship or aircraft squadron. The law does not, however, designate women as noncombatants, nor does it include any restrictions on the assignment of women to units located in or transiting combat or hostile fire zones. Women permanently assigned to a ship are to remain on board the ship regardless of where it operates. As explained by Rear Admiral James R. Hogg, Director of Military Personnel Policy, Office of the CNO, in a 1982 interview with Government Executive:

"The question is 'Do we take the women off'—the answer is 'no'. They are serving in that ship because the mission of that ship is noncombatant. The fact that the ship is going to get underway and go into a forward theater doesn't change the mission of this ship. A tender could be struck by the enemy even if it is in its home port, here in the United States."[10]

Similarly, the Navy expects women serving at shore activities to remain for the duration of their tours and would withdraw them only if other military support personnel were evacuated. Removal of personnel from a crisis or war zone is to be based on job importance, not gender.

In consonance with the law, the Secretary of the Navy policy issued in 1979 stipulates that while Navy and Marine Corps women are to have assignments "commensurate with their capabilities to the maximum extent practicable," they are not to be assigned to combat duty. The directive lists the ships authorized for the permanent assignment of women and includes auxiliaries, research ships, training ships, certain mobile logistics support force ships, and service craft. The list does not include underway replenishment ships such as oilers, ammunition ships, etc. Although these ships do not have a combat mission in the sense of seeking out and engaging the enemy, they have been considered such an integral part of the forward battle group which does seek out and engage the enemy that currently women are not to be permanently assigned to their crews. In aviation, women are eligible for permanent assignment to any billet in force support and training squadrons. This includes flying combat aircraft in noncombat missions, such as training pilots in air combat maneuvering. They also can be permanently assigned to support jobs in shore-based combat aircraft squadrons as long as the assignment does not require them to participate as crew members in aircraft with combat missions.

In terms of the Marine Corps, Secretary of the Navy policy provides that women can be permanently assigned to duty in rear-echelon billets for combat support and combat service support functions not requiring them to deploy with the assault echelon should a contingency arise. Marine Corps policy, however, stipulates that women will be assigned to support units only so long as such assignment does not routinely expose them to combat action. Furthermore, women are not to be permanently assigned, either in peacetime or during armed conflict, to units whose missions and contingency roles could reasonably be expected to routinely involve them in combat. Finally, women are not to comprise more than 10% of a unit's personnel strength. The rationale for the limitation is that in a "worst case" situation requiring the removal of all women, the unit would still have 90% of its personnel, which would be sufficient to accomplish its mission.

With regard to temporary duty, the general secretarial guidance permits the assignment of women to any ship or squadron in the Navy for a maximum of 180 days, provided the unit is not expected to have a combat mission during the period of temporary duty. If women are temporarily assigned to a ship which subsequently is assigned a combat mission, every reasonable effort is to be made to disembark them prior to execution of the mission. However, a ship's mission comes first and it would not be required to return immediately to port just to offload women. The temporary assignment of women to ships is authorized in performance of their normal military duties, including carrier-on-board missions, carrier qualification, deployment to auxiliaries as members of detachments from helicopter combat support squadrons, and for training or other professionally related purposes. In 1981, however, the temporary duty provisions were made more restrictive by the CNO: assignments will normally be performed in the Second and Third Fleets, which operate off the U.S. coasts, and temporary duty elsewhere is to occur for only relatively short periods and in special circumstances. The fleet commanders, in turn, added more restrictions, such as requiring at least two women to go on temporary duty to a ship at the same time.

The legal restrictions on the assignment of women and their amplification in policy greatly affects the employment of Navy and Marine

Corps women and women's career opportunities and progression. Because of their combat relationship, 13 enlisted skills and two officer specialties are closed to Navy women, and women Marines cannot enter four major occupational groups. Permanent assignment to all ships and aircraft squadrons with combat missions is also precluded. This effectively eliminates a major portion of Navy and Marine Corps billets as assignment possibilities for women. Also, the number of women is restricted at units which have combat support/combat service support functions and those which have responsibility for immediate fleet support or augmented fleet units in crisis or combat situations. For example, in addition to the 10% limitation on the assignment of women to the Fleet Marine Forces, the number of Navy women who can be assigned to Naval Construction Battalions and certain Naval Security Group units is limited.

The consequence of these limitations is that women are not fully assignable, and therefore the number of women the Navy and Marine Corps can effectively use is constrained. Both services have undertaken extensive research to determine the maximum number of women they can absorb. The Navy's study of enlisted women's utilization was first conducted in 1979 and has been repeated annually to revalidate the results. Key parameters of the study include the assignment constraints caused by the combat restrictions and avoiding a negative impact on the CNO's goal of three years of shore duty for every three at sea (3/3 sea/shore rotation). The study also reviews the billet structure to ensure that women have viable career paths. The results show that the Navy can effectively utilize 45,000 enlisted women and should reach that number by about 1985. It cautions, however, that attaining the 45,000 level will require almost one-third of Navy enlisted women to be in nontraditional skills.

As mentioned, an important premise of the research is that the increase in the number of women is not to have an adverse effect on male 3/3 sea/shore rotation. Despite this, during the decade male sailors often complained that "women were taking up all of the shore billets," or commanding officers ashore said they had "too many women assigned." With regard to the first complaint, while the study ensured that sufficient billets were available for men, a sailor looking for a particular job may have found it filled by a woman. Rather than focusing on the fact that the billet was filled and the incumbent's rotation date did not match his, when it was filled by a woman, he tended to generalize that "women are taking up all the shore billets." In point of fact, a shore billet was available for him somewhere as evidenced by the fact that he undoubtedly went ashore. As to the commanding officers' comment about "too many women," in reality it was usually a case of too few men. With emphasis on assigning men to sea duty, often the only personnel available for shore duty were women; this was especially true for the lower paygrades. In many cases, the choice was a woman or no one at all. In fact, women helped fill some critical shortages in shore manning. While this was generally true, the Navy did experience some problems with the assignment of a disproportionate number of women to some overseas locations. In 1981, it conducted a thorough review of the billet structure and rotation pattern for each rating and subsequently reduced the overseas requirements for women in some ratings. Additionally, in 1982 the Navy was developing a shore manning plan in an effort to more evenly distribute enlisted women throughout the shore establishment.

In 1979, the Navy also undertook studies to determine the number of women who could be utilized and offered a viable career path in each officer community. Again because of legal constraints, the number of women who could enter the warfare specialties each year was very limited—for example, 17 in surface warfare, 20–30 pilots, 15 NFOs, and 8 in special warfare. Additionally, ceilings were set on the number of women in the Supply Corps, Chaplain Corps, the Civil Engineering Corps, and all but one of the restricted line communities. In the remainder of the restricted line and staff corps, no maximum number was set, because assignment constraints were minimal and the opportunity for growth far exceeded the probable ability of Navy to recruit the women.

The Marine Corps also undertook research to estimate the number of enlisted women Marines it could utilize. Major factors affecting its study were similar to the Navy's: the number of skills and units closed to women because of combat restrictions, the need to reserve billets to ensure a satisfactory rotation base for men returning from forward-based combat units, and ensuring career promotion opportunity for women. By 1982, this study, as well as a mid-range (1985–1995) one, had not been completed.

The legal and policy constraints on the assignment of Navy women greatly influenced their careers, especially those women in the warfare communities. Women's career paths were the same as men's in terms of the billets they filled, but differed in terms of the types of units. A woman surface warfare officer looked forward to a career which included billets as division officer, department head, executive officer, and commanding officer, but only on board noncombatants. While this "separate but parallel" career progression appeared feasible in theory, women questioned its viability in practice. In surface warfare, for example, most men aspired to be executive officers as lieutenant commanders and commanding officers as commanders, but, because of the billet structure of the ships to which they were assigned, most women would not become executive officers until they were commanders and commanding officers until captain. Since most women would be department heads as lieutenants, there would be a large gap in their sea experience at the lieutenant-commander level. Another difficulty was that the temporary duty policies were overly restrictive and detrimental to women's careers. The "two-woman rule," for instance, caused great hardship on women trying to get underway time in order to qualify as surface warfare officers. Assignments to forward deployed underway replenishment ships adversely affected women in the special operations and aviation communities. For example, women in helicopter combat support squadrons were not permitted to deploy with detachments assigned to ships in the Mediterranean. Additionally, women in special operations could not enter the functional area of explosive ordnance disposal (EOD) because EOD teams normally spent six months temporary duty aboard forward deployed ammunition ships.

Because the law inhibited the ability of the services to employ women and expand opportunities for them, the Carter Administration sent a bill to Congress calling for repeal of the combat restrictions in Section 6015. Within the Department of the Navy, a split emerged between the civilian and military leaders over the proposed legislation. The civilian leaders supported repeal as a long-term solution to the management problems concerning the employment of women. Secretary of the Navy W. Graham Clayton testified during a 1978 congressional hearing that he had no objections to removal of the legal prohibitions and leaving the matter of the assignment of women to the service secretaries. He did make it clear, however, that if the law were repealed, he would continue the combat restrictions through policy. The subsequent Navy Secretary, Edward Hidalgo, maintained basically the same position. On the other hand, senior Navy and Marine Corps officers opposed both women in combat and repeal of the law. In their view, the services had sufficient manpower resources and did not need women on board combat ships or in ground or air combat units. They also questioned the ability of women to perform physically under the extreme conditions of combat. Their position was that repeal of Section 6015 had to be resolved by Congress in the broader context of the issue of women in combat. The legislation died in committee.

Unlike the other sea services, by 1982 women in the Coast Guard had no legal, regulatory, or policy constraints on their assignments. Section 6015 clearly does not apply to them in peacetime since the Coast Guard is under the jurisdiction of the Secretary of Transportation. The Coast Guard also maintains that Section 6015 would not apply to it even if it should come under the Secretary of the Navy in time of war. The Navy Secretary could conceivably establish restrictions on the assignment of Coast Guard women as a matter of policy, although even this is questioned, because it is not clear if Coast Guard ships would be Navy ships or would remain Coast Guard ships. Despite the legal ambiguities, the position of the Coast Guard is that all women are to remain on their vessels in wartime and that the assignment of women in the Coast Guard would not be changed in the event it should come under the Navy. The Coast Guard sees this issue in a very pragmatic light, maintaining that it would not be practical to remove women crew members from their vessels; to do so would weaken military readiness.

Impact of Women

In addition to the issue of combat restrictions and their effect on the utilization of women, concern was also expressed, particularly as the number of women began to increase sharply, about the impact of women on the readiness and effectiveness of their units. The issue raised most often in this context has been pregnancy. Two aspects of pregnancy cause problems for commands: unplanned losses and time away from the job. In the case of unplanned losses, the severity of the impact varies. Pregnant women at shore activities who request discharge are to give at least four months' notice in order to provide time to find a replacement. However, given that the assignment process routinely takes seven months to fill the billet, a command could have a vacant job for several months. For certain overseas locations and all ships with women assigned, the impact is more severe. Women at overseas locations which do not have adequate medical or housing facilities will be transferred to other locations with facilities. Women who become pregnant while assigned to a ship are to be reassigned to duty ashore as soon as practicable after the pregnancy is confirmed. Thus, a pregnant woman generally leaves the ship almost immediately. Since the ship has little advance notice of the woman's departure and might not receive a replacement for several months, the impact on the ship can be substantial, especially if the woman is a leading petty officer or qualified in a particular skill.

While unplanned losses due to pregnancy cause problems for commands, the aspect of the pregnancy issue which seems to receive the most attention is the amount of time women are away from their jobs as a result of pregnancy. Pregnant women miss work because of illness and having regular appointments with doctors. They can be placed in light duty and sick-in-quarters status. They are hospitalized to have the child, and are normally authorized 30 days' convalescent leave which can be extended by an additional 15 days. Generally, pregnant women are absent from their jobs for about two weeks during the pregnancy and six weeks after delivery.

There is little question that pregnancy could cause difficulties for commands with women assigned. On the other hand, the policies of all the sea services stipulate that a pregnant woman is expected to retain a high degree of commitment to the service and fulfill all her professional responsibilities. The pregnant woman is in a medical status and subject to the policies and practices prescribed by medical personnel. She is not to be excused from any duties or her job changed unless her doctor indicates that such actions are necessary. In the case of one naval regional medical center, guidance issued in 1980 stated that the normal woman with an uncomplicated pregnancy should be able to continue working eight hours a day and stand watches as long as 10 hours of rest were ensured in a 24-hour period. A woman is not entitled to any "maternity leave," she receives only the normal convalescent leave given any hospital patient. Finally, it is important to put the amount of time away from a job and the reasons for the absences in perspective. A 1978 Navy study revealed that men were absent almost twice as many days per year as women. Also, women's absences tended to be more medically-related, while men were more likely to be absent for disciplinary reasons.[11]

In 1982, a few months after becoming CNO, Admiral James D. Watkins was quoted on the pregnancy issue during an interview with *Navy Times*:

"No commanding officer that I have talked to in the last two years has told me that he feels that the pregnant woman issue is detrimental to the overall readiness of his command.

" . . . when I ask him in a forthright away, 'Is this woman doing a good job?' I have invariably received the answer that she is doing a superb job. Then, I say, 'Is her lost time, including pregnancy, more detrimental than the lost time of a male?,' and the answer is 'No.' She has never had any lost time until she was pregnant, so she banks it up, if you will, by good behavior, and not going on unauthorized absence, and not deserting, and not involved in some of the immature male endeavors that are characteristic of a lot of our youngest and immature sailors."[12]

A second area of concern was the impact of military couples. As the number of women in the sea services rose, the number of couples in which both members were military also increased. The most specific, immediate difficulties relate to shipboard duty. If the couple is assigned to the same ship when they marry, one of them must be reassigned since appropriate berthing accommodations are not available and conflicts in their military relationship may arise. Which individual is transferred varies, depending on the skills, projected rotation dates, and the needs of the command. Regardless of which one is reassigned, it is an unplanned loss for the ship. If a couple marries while assigned to different ships, again one of them is transferred, if the couple does not have any dependents. The rationale in this case is that, since each is still considered single for housing allowance (BAQ) purposes and thus would not draw the money while on sea duty, the Navy does not want the couple to suffer the undue economic hardship of losing all of its BAQ. The couple can, however, volunteer to remain on sea duty.

In overall terms, the policies of the sea services are to try to assign military couples to the same location, but they do not guarantee such assignment. Generally, throughout the period from 1972 to 1982, couples were colocated, but the difficulty of doing so was most severe in the Coast Guard because it had so many small units scattered at widely separated geographic locations. In some respects, the success of the services in arranging colocations caused problems, because couples assumed they would be assigned together. The services came under increasing pressure to change rotation dates or normal assignment patterns for at least one member of the couple in order to colocate them.

Another aspect of the problem arose if the couple had children.[13] Aside from the immediate problems associated with any dual-career family, such as finding adequate day-care facilities, the services were faced with unique problems. For instance, in a crisis at an overseas location, the couple would be required to remain at their duty stations but their children would be evacuated. Because these problems were becoming more acute, in the early 1980s, the Navy began to develop a dependent care plan requiring the couple to specify in writing who would take care of their dependents in the event they were unable to do so because of military duties.

Child care and special assignments also raised concern with regard to parents without partners. Although in 1980 the Navy had more male single parents than female by about two to one, the proportion of women who were single parents was higher (about 5% versus 1%). As a result, as the number of women increased, the number of single parents rose. Nevertheless, Navy policy basically treats single parents the same as other Navy personnel. They are not to receive preferential treatment in assignments and are expected to carry out all of their professional duties. They can request a hardship discharge, but they must meet all criteria for the discharge and will not be separated from the service simply because they are single parents. If they fail to meet their military responsibilities, they can be involuntarily separated. Overall, most commanders did not consider single parents to be a major problem, but some units, such as those with rotating watches or at certain overseas locations, had difficulties. To help alleviate the situation in 1982 the Navy was planning to require single parents to have dependent care plans similar to those of military couples. The requirements are expected to be published this year.

Another area of concern involved fraternization. During the decade, the sea services continued to rely on custom, tradition, and the judgment of local unit commanders to govern both professional and personal relationships among their personnel. They also relied on general regulations which called for all personnel to contribute to the "good order and discipline" of the command. The Navy did, however, provide some specific guidance when women were first assigned to ships. In a 1978 memorandum, the CNO stated that men and women on board ship were considered to be in a duty status at all times and that conduct was expected to meet traditional standards of military decorum. He also indicated that overt displays of affection were out of place and should not be permitted. Consequently, some commanding officers of ships issued specific regulations regarding fraternization and public displays of affection. Generally, however, the sea services relief on custom and tradition rather than explicit service-wide rules and regulations to minimize fraternization. Overall, the sea services did not consider this to be a major problem, although the Marine Corps expressed greater concern than the others.

Closely related to fraternization was the issue of sexual harassment. Although the sea services did not have exact data on the incidence of sexual harassment of military women, indications were that it was substantial. Apparently, however, it was not any worse than for civilian women working for the federal government or American society as a whole. Nevertheless, the services recognized that sexual harassment hurt the professional development and performance of women and thereby adversely affected services as a whole. Beginning in the early 1980s, they issued service-wide policy guidance and regulations which defined sexual harassment, emphasized its unacceptability and outlined procedures for reporting incidents and actions to be taken. They also started training programs to assist commanders in dealing with the problem.

Finally, another major area of concern is the ability of women to perform physically demanding nontraditional jobs. In 1978 the Navy began a five-year study on occupational physical standards. While the increasing number of women and their new assignments provided the impetus for the study, the study itself is gender-free, that is, the results are to apply to both men and women. The objectives of the study are to identify muscularly demanding tasks, determine the percentage of men and women capable of performing the tasks, and validate a strength test battery as a predictor of ability to accomplish the tasks. The test battery is designed to measure an individual's ability to lift a particular weight, to carry and walk or run with the weight, and to push, pull, squeeze, turn, or swing objects. Developing the criteria for muscularly demanding tasks and applying task performance tests are being done under actual working conditions on board ship and in squadrons. Additionally, the study is reviewing the ability of women to perform certain general tasks aboard ship, such as extricating injured persons, operating a fire hose, and exiting watertight doors and scuttles. While the Marine Corps has not undertaken any specific research of its own, it is monitoring the Navy's study.

In addition to looking at the physical ability of individuals to perform tasks, the Navy and Coast Guard also have studied shipboard equipment, fittings, and protective clothing to determine the impact of their design on the ability of personnel to use them. Generally, women did have greater problems, but smaller men had difficulties as well. On the other hand, women actually had fewer problems than men in some areas, such as exiting through small hatches. Additionally, the Navy began to include women in its sample populations for various research projects involving human engineering, such as a study on body dimensions and shapes used to size and configure aircraft and equipment.

While all of the areas discussed above affected them to varying degrees, generally the sea services viewed them as management problems that could be solved. Furthermore, the bottom line was that during the decade women performed well and contributed to operational readiness. Navy leaders praised the performance of women both in general and with regard to operational forces. In a 1982 speech, Deputy Assistant Secretary of the Navy for Manpower E. C. Grayson characterized the performance of Navy women as exceptional.[14] Admiral Watkins in his 1982 interview with Navy Times stated that women were generally performing better than men.[15] Vice Admiral Lando W. Zech in his testimony as Chief of Naval Personnel on the fiscal year 1983 budget offered the opinion that: "Overall, women continued to contribute to the operational effectiveness of the Navy by skillfully and confidently performing their duties both at sea and ashore."[16] Within the operational forces, Vice Admiral Thomas J. Kilcline, Commander, Naval Air Force Atlantic Fleet stated in a 1982 report to the CNO that women in his units appeared to be better educated, performed as well, advanced more rapidly, and presented fewer disciplinary problems than men.[17] With regard to shipboard duty, a 1981 Navy-wide message from the CNO declared: "Without question, the women in ships program has been an impressive success. Women are routinely performing in both traditional and non-traditional areas with skill, confidence, and dedication."[18] Commanding officers of ships with women assigned reported that women had proven themselves capable of handling shipboard tasks, including long hours and hard work. The general perception was that women were able to perform routine shipboard jobs better than anticipated. In some cases this was the result of women's greater motivation and in others of their higher educational levels.

The view of Coast Guard leaders echoed that of the Navy. In 1978, Rear Admiral William H. Stewart as Chief, Office of Personnel in the Coast Guard testified to Congress that women had made and were continuing to make a significant contribution to the operational readiness of the Coast Guard. In terms of shipboard duty, he indicated that their performance had been outstanding and that he personally, as a former commanding officer under combat conditions, would have no hesitation in having them on board ship with him.[19] The commanding officer of the first ship to have a mixed crew, Captain Joseph McDonough, considered women to be an asset. As related by Rear Admiral Paul Yost, Chief of Staff of the Coast Guard, "Whenever he [Captain McDonough] ran into fellow commanding officers with all male crews he would invariably receive their condolences. 'No,' he said, 'You have my sympathy . . . with women on board, *Morganthau* is a better ship for it.'"[20]

Generally, the Marine Corps was less vocal publicly about its assessment of women Marines. Lieutenant General Edward J. Bronars, Deputy Chief of Staff for Manpower, did state in 1979 congressional testimony that: "Our experience is that women contribute in a major way to the overall Marine Corps ability to satisfy personnel, skill, and leadership requirements. Women have become an integral and valuable part of the Marine Corps . . ."[21] In a 1980 letter to all general officers, commanding officers, and officers in charge, General Robert H. Barrow, Commandant of the Marine Corps, asserted that women Marines were an integral and vital part of the Corps and that the past contributions and dedication of women Marines spoke for themselves. He also recognized the past contributions to mission accomplishment that women Marines had made and continue to make within the Corps.[22] While the Marine Corps acknowledged that women generally performed well, it expressed great concern about the role they could fill in the Fleet Marine Forces.

Attitude toward Women

While the services had at least overtly supported progress for military women and, as has been recounted, major changes did occur, Navy and Marine Corps women began to sense a change in atmosphere as they entered the 1980s. They became very concerned that, at a minimum, they were no longer marching forward and in fact might even be forced to do an about face and march back.

The immediate cause for the concern was the election of a new administration in November 1980. President Ronald Reagan is a political conservative who did not support the Equal Rights Amendment and who, upon taking office, appointed few women to high-level positions. Within three months of taking office, the new Deputy Secretary of Defense directed the services to conduct an assessment of the impact of women on readiness and mission effectiveness. Although the major impetus for the study was the Army's earlier announcement that it would "pause" at 65,000 enlisted women while it studied various policies and practices regarding their utilization, the implication was that women were a problem in all of the services. Additionally, the Reagan Administration decided not to resubmit to Congress legislation which would have lifted the combat restrictions for women. Other actions between 1980 and 1982 which were also perceived by women as negative signals included:

- The Air Force indicated it wanted to reduce substantially its recruiting goals for women, saying that it would need far fewer women in future years than was projected in the Carter Administration.
- The Marine Corps began a review of the assignment of women to the Fleet Marine Forces, and officials said there was a possibility of reductions.
- The Supreme Court upheld the constitutionality of a law exempting women from draft registration.
- As a result of the passage of DOPMA, a new promotion system was instituted for women which potentially could jeopardize their promotion opportunities.
- Secretary of the Navy John Lehman made a speech to a group of women officers in which he appeared insensitive to their concerns, and showed a lack of knowledge of the career paths.

The impact of these actions and the media's coverage of them, according to a report to the Secretary of Defense, by the Defense Advisory Committee on Women in the Services (DACOWITS), was to lower the morale of women in the military and contribute to "an underlying uneasiness among service women about their perceived value to the military and the implied assumption that the presence of women denigrates readiness."[23]

Lastly was the continuing problem of the attitude of individual men toward women as they worked together. As noted by Admiral Zumwalt in a 1982 interview with *The Times Magazine*, during his tenure as CNO he found that, "Although the Navy was a racist institution, it was far more of a sexist institution. I found it easier to deal with racism than sexism. It takes longer for a white man to come to believe that a white woman is his equal than it does for him to come to believe that a black man is his equal."[24] Navy and Marine Corps policy certainly did not condone such an attitude, but as the decade passed, women noticed that it seemed increasingly "okay" to both have and voice a biased attitude.

Even though troubled by the changing institution atmosphere, women noted some hopeful signs. Obviously progress had been made over the decade. Also, the five-year Navy Affirmative Action Plan issued in late 1981 dealt with a number of the issues which concerned women, such as the promotion opportunity and improving leadership training for a "mixed gender" work place. In the case of the Marine Corps, in late 1980 the Commandant issued a policy statement about leadership and responsibilities pertaining to women: "Women Marines and male Marines serve side by side in our ranks. They are equal in every sense. They are Marines. They deserve nothing less than outstanding leadership, equal treatment, and equal opportunity for professional development."[25] Furthermore, in early 1982, the Office of the Secretary of Defense issued several memoranda emphasizing the contributions of military women and instructing the services to ensure that women are not subject to discrimination in career opportunities. Lawrence J. Korb, Assistant Secretary of Defense for Manpower, Reserve Affairs, and Logistics, spoke to a group of women officers to reassure them that the outlook for military women was good. While most women still felt there was much room for improvement in attitudes and leadership, the mood could be characterized in 1982 as one of "cautious optimism."

The Future

The changes which will occur between 1982 and 1992 probably will not be as dramatic as those which took place between 1972 and 1982. By 1982, all restrictions had been removed for Coast Guard women and, except for Section 6015, most major institutional barriers had been eliminated for Navy and Marine Corps women. The next decade will see more women in the fields already open to them, but women generally will not be found in many new or different skills. The one factor that could result in a dramatic change, however, would be a court ruling permitting or perhaps even requiring, the assignment of Navy and Marine Corps women to combatant units. Meanwhile, the number of women will continue to increase because women will perform well and contribute to the operational effectiveness of the services and because the number of men available and willing to join the military will decline. Pragmatism will overcome institutional reluctance. As a Coast Guard paper on women observed, "We need them."

Notes
1. While most, if not all, of the changes in law and military policy probably would have occurred eventually, the legal challenges accelerated the process. In the case of pregnancy discharges, suits brought by a Navy enlisted woman, a woman Air Force officer, and a woman Marine all helped to convince the services that they were better off changing the policy rather than losing a court case. With regard to the dependency requirement for military women, the case brought by First Lieutenant Sharron Frontiero, USAF, led directly to a change in the law. In terms of not admitting women to the service academies, Congress certainly had to be influenced in its deliberations by the fact that suits were pending against the Navy and Air Force which had been filed by two women who wanted to attend those academies and by four members of Congress who objected to the discrimination against women. Finally, while Congress was considering amending the law governing the assignment of Navy women, Judge John J. Sirica ruled in a case brought by seven Navy women that the law was unconstitutional and the Navy could not automatically exclude all women from sea duty. The amendment subsequently passed by Congress met the court's guidance.
2. All data are as of the end of the fiscal year, except data on marital and parental status, which are as of March 1980.
3. The General Unrestricted Line consists of officers with the 110X designator. About 75% of the community is women, because men in the unrestricted line normally are in the warfare communities. Men are not recruited into the 110X community, and it is not considered to offer a viable career path for men. Most of the junior officers in the community are surface warfare trainees (116X) who have not qualified as surface warfare officers in the appropriate period of time. Among the senior officers are the 110Xs who did not convert to surface warfare when the community was formed and those who have a particular expertise the Navy needs.

4. The data shown represent cohort attrition for the first three years of service, that is, the percentage of enlistees who entered the service in a given fiscal year who did not complete the first three years of their enlistment.

5. The data shown represent net retention, that is, the percentage of enlisted personnel eligible to reenlist who did reenlist.

6. On the career side, during the decade Navy women always had a lower reenlistment rate than men. The striking feature, however, is that the rate was consistently parallel to men's, indicating that factors influencing men to stay in or get out have about the same impact on women. In the case of the Marine Corps, reenlistment rates for women varied over the years compared with male rates. This lack of consistency resulted at least partially from the fact that the number of career women eligible for reenlistment was still relatively small, so a small shift in the number of reenlistments had a large effect on the reenlistment rate. Much the same situation prevailed in the Coast Guard. For example, the extremely low career reenlistment rate of 21% for women in 1982 was based on only 14 women reaching the end of their enlistments.

7. Despite the stated stringency of the policy, a few exceptions were granted, generally to women who were nearing the 20-year retirement point.

8. U.S., Congress, House, Subcommittee No. 2 of the Committee on Armed Services, *Hearings on H.R. 9832 To Eliminate Discrimination Based on Sex with Respect to the Appointment and Admission of Persons to the Service Academies*, 93rd Cong., 2nd sess., 1974, pp. 89–90.

9. *Frontiero v. Richardson*, 411 U.S. 677 (1973).

10. Barbara J. Oganesoff, "Women in the Military: Part II, The Human Power Issue: From Policy, to Utilization Models to Field Operations," *Government Executive*, March–April 1982, p. 48.

11. Marsha S. Olson and Susan S. Stumpf, "Pregnancy in the Navy: Impact on Absenteeism, Attrition, and Workgroup Morale," Navy Personnel Research and Development Center Technical Report 78–35, September 1978.

12. Rick Maze, "Need for Women at Sea to Grow, But Not in New or Different Roles," *Navy Times*, 25 October 1982, pp. 12, 20.

13. In 1980, of Navy women married to civilians, 31% had children. Among dual military couples, 21% had children. While this was a major increase, the proportion of military women who were parents was still considerably lower than for men. For example, 70% of military men with civilian wives were parents. Also, military women tended to have fewer children.

14. Speech given by Mr. Ellison C. Grayson, Deputy Assistant Secretary of the Navy (Manpower) to the Defense Advisory Committee on Women in the Services (DACOWUS), San Diego, Ca., 26 April 1982.

15. Maze, *Navy Times*.
16. U.S., Congress, House, Committee on Armed Services, *Hearings on Military Posture and H.R. 5965 Department of Defense Authorization for Appropriations for Fiscal Year 1983*, Part 7, 97th Cong., 2nd sess., 1982, p. 112.
17. Rosemary Purcell, "Pregnancy Hurts Navy's Readiness: Kilcline," *Navy Times*, 5 July 1982, p. 14.
18. "Assignment of Women to Shipboard Duty," NAVOP 001/81, CNO msg 022240Z Jan 81.
19. U.S., Congress, House, Military Personnel Subcommittee of the Committee on Armed Services, *Hearings on Women in the Military*, 1979–1980, pp. 157–58.
20. U.S., Coast Guard, "RADM Paul A. Yost: 'Women in My Work Force,'" *Commandant's Bulletin*, 45–81.
21. Hearings on Women in the Military, p. 144.
22. "Leadership and Responsibilities Pertaining to Women Marines," White Letter No. 18–80 from the Commandant of the Marine Corps, 2 December 1980.
23. As quoted in Richard Halloran, "Military Told to Lower Barriers to Women's Service," *The New York Times*, 31 January 1982, p. 25.
24. Jay Finegan, "The Man Who Changed the Navy," *The Times Magazine*, March 1982, p. 34.
25. White Letter No. 18–80.

AUTHOR'S NOTE: *The author gratefully acknowledges the contributions of those who provided data for this article: Commander Joyce Kilmer, USN, and her staff; Lieutenant Colonel Ruth Woidyla, USMC; Lieutenant Judith Hammond, USCG; and Lieutenant Colonel Harry Thie, USA.*

"Women in a Changing Military"

8

Edna J. Hunter and Carol B. Million

U.S. Naval Institute *Proceedings*
(July 1977): 51–58

Tradition has usually granted women the prerogative to change their minds; society has been extremely reluctant to allow them to change their roles. And within the military system, a traditionally male-oriented social institution, women's roles have been rigidly circumscribed—until recently.

THERE ALWAYS HAVE been women involved with the military, either formally or informally. Although women have been only tolerated in organized military situations, they have always been an integral part of guerrilla forces where the need for "people power" is acute. For example, the *vivanderas* of the Mexican Revolution were heralded for their abilities to overcome adversity, as were women like Ho Tzuchen (third wife of Chairman Mao Tse-tung) who were involved in the Chinese civil war of the 1930s. It is not unexpected that many of the main characters in underground militant activities are female.

In crisis situations, traditional role definitions of women's "proper" places are cast aside. Even in the more traditional structures of the military, during emergencies when manpower is in short supply, women have assumed roles that in peacetime are exclusively the prerogative of men. During military crises, women are pressed into service, either as active duty military personnel or as civilians a la Rosie the Riveter of World War II.

The United States is not now involved in a military emergency, but since the end of the draft, the military has gotten into a people-power squeeze in which the services are actively competing for *qualified* individuals. Women are now viewed as an untapped volunteer source, and a renewed interest by the military in the capabilities of women as military members is to be expected. With about 109,000 women on active duty in the services today, and with the percentage of women increasing, it behooves the military establishment and those involved in research into the needs of the military to look closely at what women can do for the services and what the services will have to do to attract qualified women in the future. Women are no longer ancillary to the services, to be tolerated and dispensed with as soon as possible.

History of Women in the Military: The usual pattern for women in the U.S. military services has been a call to duty during a time of need and a dissolution of the service after the crisis subsides. Such a pattern makes it impossible for the servicewoman to obtain military experience in a variety of assignments which would ensure promotion within the organization.

During the World War I era (1917–1920), the Navy used the yeoman (F) rating to free men from administrative duties and send them to sea. More than 13,000 women served in the Navy and Marine Corps. Many others probably would have served in the Army, except for a law then in existence which required the Army to enlist only "male persons," whereas the Navy could enlist "citizens" (perhaps a term that was interpreted rather loosely in the case of women because they were generally not permitted to vote until 1920). However, necessity is the mother of rule-bending, and the Navy was able to overcome this handicap, as well as a rule requiring that all yeomen be assigned to a ship. Then as now, women could not actually serve in ships, so the yeomen (F) were assigned to a boat sunk in the mud of the Potomac River.

The next World War brought the beginning of women's services as they are today. Congresswoman Edith Rogers sponsored a bill which became law in May 1942 and resulted in the organization of the Women's Army Auxiliary Corps (WAAC). This group did not get full military status, although two months later the WAVES (Women Accepted for Voluntary Emergency Service) were designated as military persons, not auxiliaries. In 1943, the Army followed suit, and the Army women became WACs— in the Army, not with it. Later women's corps in both the Marine Corps and the Coast Guard (SPAR) were also formed. Military women did a variety of jobs during World War II, although almost half were in administration. In 1945, WAVES represented 55% of the uniformed Navy personnel in Washington, D.C.

Although women had been incorporated into the services merely to fill a wartime need, President Harry Truman signed the Women's Armed Services Integration Act of 1948 (Public Law 625) by which women were authorized to become an integral part of the services. Women, however, still found themselves restricted in ways men were not:

- Women were required to be older than men to enlist (18 as compared to 17) and had to have written parental consent if under 21 (as compared to 18 for men).
- Women could not exceed 2% of total regular military enlistment strength, and female officers were not to exceed 10% of the female enlisted strength.
- Women officers could not have permanent commissions above the rank of commander (Navy) or lieutenant colonel (Army, Air Force, and Marine Corps).

Although most of these rules have now been abolished, either by new policies of the services or by court cases contesting their constitutionality, they held true for U.S. servicewomen from 1948 until 1967—

the year in which the 2% ceiling on military women was lifted because of the escalating Vietnam conflict. Today, the number of women in a particular service is determined by the service secretary. The Air Force and Army have established the highest quotas at 15% and 6% respectively; the Navy is at 4.5% and the Marine Corps, 1.5%, although these percentages continue to change.

In the early Seventies, the status of women in military service changed drastically. Two landmark changes bear noting . A legal decision in 1973, *Frontiero v. Richardson*, greatly changed the legal rights of military women and their children. One year previously, in 1972, Admiral Elmo Zumwalt, Jr., Chief of Naval Operations, promulgated Z-Gram 116 which augmented considerably the status of women in the Navy. These changes included:

- Entry of women into all enlisted ratings but in limited numbers
- Suspension of restrictions on women succeeding to command ashore
- Chaplain and Civil Engineer Corps opened to women, thereby allowing women officers in all staff corps
- NROTC, including scholarships, opened to women
- Women permitted to achieve flag rank within managerial and technical specialities

These formal changes in military structure have been further facilitated by the end of the draft and by the reemergence of the women's liberation movement which "legitimized" the use of women in nontraditional jobs and made them more willing to accept such jobs.

Profile of the Military Woman: Who are the women who join the military services? Demographically, they are very similar to the men who volunteer. A typical female enlistee is 20 years old. She must be between 18 (17 with parental consent) and 26; males average slightly younger than female enlistees. She is probably a high school graduate with no college or other post-high school training. Her male counterpart is less likely to be a high school graduate. She is probably from a small town and from an intact family with two or three brothers or sisters. Her family is not particularly opposed to her choice of the military, and her friends are about equally divided between supporting her decision and denigrating it. She knew little about the service before entering it and had no female friends who had ever been in the service. This is in direct contrast to the male recruit who probably has had friends, a father, or a brother who were in the service and could thus give him information about what to expect. The lack of information for the female enlistee is not ameliorated by recruiters who know little or nothing about the options open to women in the military services. Lack of realistic information about military life is a severe problem for the woman enlistee because it frequently leads to unrealistic expectations. When the expectations are not fulfilled, she experiences disillusionment and dissatisfaction with the actual situation. This may in part explain the higher turnover rate of female enlistees in past years. However, new regulations are making early discharge more difficult.

Although the female recruit may not have much information upon which to base her decision to enter the service, typically she scores higher on the general classification test (GCT) than does her male counterpart. In fact, women must actually make higher scores than males on pre-enlistment tests in order to qualify for military service. Since the women who enlist are typically older than the men who enlist, it is not surprising that over 70% of female enlistees have worked before entering military service. In fact, prior work experience is almost assured since women must wait approximately six months between the time they sign up and the time they report for basic training. There are more female volunteers than the military can use—in direct contrast to male volunteers who are in such short supply that recruiters are hard pressed to meet their quotas. In the past, women were "true volunteers" since, unlike men, they volunteered without any compulsion from the draft or psychological pressure. Even so, they suffered from a quota system. Not all those who wanted to volunteer were able to. Ironically, those who couldn't enlist missed out on such things as veterans' benefits which have been enjoyed by men— even those drafted against their will.

Why do women volunteer? What motivates them to join organizations that are traditionally bastions of male solidarity? The overwhelming evidence seems to suggest that these women are not "radical" types. They are not ideologists. If anything, they tend to be rather "traditional" in their perceptions of women's roles in society. Although 70% have worked prior to service entry, it is noteworthy that the vast majority have held positions in four stereotypically feminine categories: office, hospital, restaurant, and factory.

Motivation to Enlist: When recruits were questioned about their motivations to enter the service, the majority (62% of women, 60% of men) list the acquisition of a new job skill as a primary reason for enlistment. Other reasons for enlistment given by both men and women are their perceptions that military supervisors care about workers' problems and that recruits feel that they are "really doing something important." Job security is also important to both men and women; however, it appears to be *more* important to men than to women. The latter may be a reflection of the economic situation. Recruits see the military as a safe haven from unemployment, or perhaps the women perceive that jobs are only a temporary part of their life pattern. (This, again, is a reflection of their conservative view of the role of women, i.e., that they will marry, have children, and drop out of the labor force. It is a view that is widely held but is probably unsupportable in view of the increasing number of working mothers). In addition, reports of female workers show they tend to value a "clean, cheerful" work environment. This desire would, of course, be incompatible with a nontraditional billet, and, thus, may account for some of the reluctance of female recruits to assume those positions.

Occupations of Women within the Services: As previously mentioned, most job classifications in the services are now open to women.

For example, the Army identifies a total of 451 military occupational specialities (MOS) and of these, 415 are open to women. The Navy has designated jobs into three categories: those *open* to women, those *controlled* as to the number of women who may enter the classification, and those *closed* to women. The vast majority of job classifications are open to women. Nevertheless, women in the services are still overwhelmingly in traditional occupations. A full 76% are employed in medical, administration, communications, and supply fields. Several explanations of this situation have been advanced. A simple reason for part of the concentration of women in very few career fields may be that many of them went into the service before other fields were open to women. Also, some women may have gone into the service already skilled in a traditional female occupation (e.g., nursing). Another reason for the concentration of women in traditional fields may be the aforementioned conservative life expectations of female enlistees. They simply don't apply for "male" jobs. Even in nontraditional occupations, women are assigned traditional "female" work; men do the job of the unit, and women do paperwork. When this situation occurs, morale lags. Women feel they aren't doing what they were trained to do, and the men feel that they are forced to do extra work to make up for jobs women aren't doing. This problem can be eliminated by supervisory persons who use female workers as an integral part of the work group, not as a special segment to be set off for special (privileged/discriminatory) treatment. There is yet another possible explanation for this job clumping within the Navy. Job classifications that are listed as "open" are not always, or even often, open to women, because few women will be used to fill jobs that are used as shoreside rotation billets for men.

In the Army the same situation occurs in that women cannot be too heavily concentrated in units that may be suddenly shifted forward to a fighting front in wartime because, under existing regulations, any women in these units must be replaced before the units go into combat. Thus, a unit with many women would be too shorthanded to go into combat without overtaxing male personnel or getting trained replacements for female members, thereby wasting valuable time. In view of the fact that few women will be used in these job classifications, it is not surprising that women do not apply to training schools for positions in which they will never be able to actually work and for which they will have to wait months simply to get into a school. Thus, they consider it better to take a traditional job for which training and job placement are assured.

Legal Restrictions Placed on Women in the Military: To understand the legal restrictions placed on military women, or women in any professional situation, we must first examine the 14th Amendment to the Constitution, specifically, the equal protection clause. A full discussion of the topic, however is far beyond the scope of this article. Nonetheless, let us briefly consider the ways the courts have interpreted this amendment with respect to women.

Where does sex fit in with the interpretation of the 14th Amendment? Three recent court decisions are noteworthy. The case of *Reed v. Reed* (404 U.S. 71; 1971) dealt with an Idaho statute that preferred male over female administrators of an estate in probate. The state's rationale for this law was that men had more experience with money. The court held that this rationale was not a "legitimate government reason," and, therefore, was not related to any appropriate government purpose. Thus, a hearing was required to determine whether a male or female would be better qualified to administer the estate.

A second case, and one that more specifically concerns us, is *Frontiero v. Richardson* (411 U.S. 677; 1973). It dealt with a married servicewoman who was denied the right to claim her spouse as a dependent for the purpose of obtaining increased quarters allowances and medical benefits for him. The rule by the Air Force was that men could claim wives as dependents regardless of the wives' incomes; however, a woman had to *prove* that she provided more than one-half of her husband's support. The rationale was that it is socially usual for men to support wives, and it would be administratively inconvenient to make all men in the military prove that they provide more than 50% of their wives' incomes. The court held that classifications based on sex, like those based on race and national origin, are *inherently suspect*. A "rational government reason" of administrative inconvenience was not good enough to stand in the face of the suspect classification. This holding seems to suggest that any classification based on sex would be suspect and carry a very heavy burden of justification.

However, in 1974, the court in *Kahn v. Shevin* (42 U.S.L.W. 4591; 1974), a case contesting the validity of a Florida statute which gave a $500.00 property tax break to widows, but denied the break to widowers, held that sex was *not* a suspect classification. It decreed that this unequal treatment of men and women was supported by a "legitimate government purpose" to help needy widows, and therefore did *not* violate the equal protection clause of the 14th Amendment. So the court again came back to the "rationally related to a *legitimate government purpose*" test to determine if classification by sex is unconstitutional.

Thus, laws classifying by sex have been differently handled by the court in several recent cases, and no clear standard of constitutionality has been established. It is uncertain how the recent rulings by the court will affect Title 10, sections 6015 and 8549 of the U.S. Code which restrict Navy and Air Force women from serving on fighting planes or on board any ships except hospital ships and transports. However, the present court is unlikely to overturn the statutes, especially in view of the deference traditionally paid by the court to the military (*Orloff v. Willoughby*, 345 U.S. 87; 1952). There is no more legitimate government function than national defense, and any rule dealing with the military, no matter how sexist, will be viewed by the court as reasonable and rational unless that clearly is not the case. Thus, classification by sex by the services, if to promote national defense, is rationally related to a legitimate government end and would be considered *not* violative of the equal protection clause. For example, in the *Frontiero* case, the court went to some pains to point out that allowing husbands of servicewomen to get dependent status did not impede the efficiency of the services in carrying out their military functions and was *not* related to a legitimate government end.

Until very recently, the automatic discharge of a servicewoman who became pregnant or adopted minor children was justified in federal courts with the rationale that pregnancy and motherhood would affect the efficiency with which a servicewoman performed her job, and that national defense might be impaired. Thus, a legitimate government need of national defense for the automatic discharge rule was established (*Struck v. Secretary of Defense*; 409 U.S. 1071; 1972). So a pregnant woman in the service was not treated as any other service person with a temporary disability. Instead, she was discharged—without the usual severance pay that service personnel given medical discharges are generally entitled to. This ruling has been recently overturned. Pregnant women may now opt to remain in the service. But it must be remembered that the present court has not clearly made sex *a suspect classification*, and any "legitimate government purpose," such as national defense, will support a distinction based on the sex of those classified. Women in the service are particularly hampered by this line of reasoning and by the great burden of justification it places on anyone trying to change the rule to prove that there is *no* legitimate government reason for the rule. This, of course, will be greatly altered by the passage of the Equal Rights Amendment, which would, in effect, make any classification by sex "inherently suspect" and subject the *service,* not those trying to change the rule, to a great burden of justification.

Problems: A major problem with servicewomen in past years was that many did not finish even three years of the four-year enlistment. A primary reason for this rate of attrition has been the fact that women may obtain voluntary discharges if they become pregnant. This type of discharge is supported by a traditional view of a woman's career as terminable whenever she becomes involved in her primary functions—wife and mother. The case of obtaining discharges reinforces the viewpoint that women are in the service only until they marry and are not really serious about pursuing careers. This view is held by many in the male military establishment. The lack of social stigma attached to an honorable pregnancy discharge for a female enlistee and the acquisition of a status that is perceived as highly desirable, that of wife and mother, are powerful influencing factors for young female enlistees to opt to leave the military.

However, the law may eventually change this situation, making it more difficult for pregnant servicewomen to get discharges, even if they request them. The courts have already ruled that the military may not discharge a pregnant servicewoman against her will. The advent of the Equal Rights Amendment would probably affect the pregnancy discharge by either allowing men a fatherhood discharge (fatherhood deferments from the draft were available for many years during the Fifties and Sixties), or by treating pregnancy as a temporary disability (as is now the case for those women who choose to remain in the service subsequent to childbirth). In either case, if female enlistees are to be maintained within the service for their full enlistment periods, it is necessary that women with a less "traditional" orientation be encouraged to join the service; that the seriousness and duration of the enlistment commitment be made clear to new enlistees; and flexible work/parenthood arrangements be available within the service so that women, and men, feel they can be adequate parents and still maintain a service career. In addition, service careers for women that are seen as valuable in their own right—not merely as holding patterns—might very well increase retention rates.

We should mention that psychiatric discharges are much more common among women than among men, either because more women have psychiatric problems, or because they are more likely to be offered such discharges on the grounds of "unsuitability" for military duty by psychiatrists who view women stereotypically. As already noted, female recruits have very little realistic information about the military system or military life, in direct contrast to male recruits who have considerably more information about what it's like to be in the service. Thus, while men may not like the service, they are more likely to know what to expect, and what is expected of them; women, on the other hand, often do not. Recruiting advertisements usually show service personnel performing interesting jobs in exotic locations but give little useful information about the service. Thus, a prospective servicewoman, more so than a serviceman, enters the enlistment contract with unrealistic expectations. Disillusionment is almost inevitable, and with it dissatisfaction with life in the military. The dissatisfaction is compounded when women learn that, because of service regulations, open billets often are not really open to women. The "exotic" job in most instances turns out to be a traditional office or hospital position. Further, even should the woman request a school to train for an unusual job, she may be unable to qualify for the position since the aptitude test used to evaluate enlistees for positions in training schools measures mechanical skills, such as those gained in high school shop classes that women seldom are allowed to take. A low test score by a female enlistee would, thus, preclude admittance to a technical school, thereby leaving women in fields such as typing that they are already trained for. Even where the woman is able to get into a school, successfully complete the course, and attain a skill in a specific area, she frequently is not placed in a billet requiring that skill. Rather, she finds herself in a lower-status position because of the restrictions on women with respect to combat assignments and the fact that certain billets are designated specifically for ship-shore rotation; thus, they are closed to her. Her full potential is not utilized and, thus, her dissatisfaction with the service increases.

Because of the conservative structure of the military and because of the traditional nature of women who join the service, job innovation is rare. Those individuals who would change the system find it easier to leave it; those who find their present traditional roles rewarding, stay. Most military women have few models of female military superiors to whom they may look for guidance. Without guidance as to what an enlistee can be and what she can do, the woman becomes discouraged and leaves, perpetrating the belief that women are unable to handle military life. It must be remembered that her military experience is much different from that of her male counterpart; he has role models to aspire to and more adequate information about the military throughout his career. He is not afforded an easy option to leave the service prior to the expiration of his enlistment.

Attitudes of Military Men toward Servicewomen: In any work group, the attitudes of coworkers and superiors are important in determining the amount of success one has in completing assigned tasks. In a military group this factor is compounded by the length of time a work group spends together, often longer than the usual workday. One's coworker may also take on added significance because of the lack of support from non-service friends or family nearby. Relationships with one's coworkers become crucial to military women. The way women are treated by the men with whom they serve may actually affect job performance.

Men may fear that women will decrease unit effectiveness. Will women be able to do heavy work? How will the presence of women in previously all-male work groups affect the men? Experience has shown that the first women going into a unit that was previously all-male are frequently given the peripheral "feminizable" jobs of the unit. For example, a trucking unit whose primary job is to ship supplies will assign women as dispatchers or have them check bills of lading. Depending on the desirability of these jobs, the women may either be resented because they are assigned easy jobs while the men have to "really work," or they may be tolerated because they are performing a dreary, unpleasant task. In either case, the women are not really involved in making a valuable contribution to the primary purpose of the unit. They are still "auxiliary" and not an integral part of the process of the unit. Further, they are not using their training. As time passes, either because of pressure from the top to integrate units, or because there are too many women to let them remain on the periphery, women become involved in the actual business of the unit. At that point, competition between male and female coworkers may arise. If competition is reasonably and fairly handled by superiors, it may give rise to greater production by the unit and even to mutual cooperation, as men find that women behave much as the men do, and that women are able to hold up their end of the job. If competition is improperly handled by superiors who have preconceived notions about women's proper roles, women in the unit will become increasingly discouraged as it becomes clear they are in a "no win" situation. They do not have equal footing with men, and because of the job requirements, they cannot act in traditional feminine roles. Their only alternative, then, is to leave.

Men in the unit may also resent what they perceive as "special breaks" that the women are receiving. However, as male supervisors become more accustomed to women workers, and as women work their way into supervisory positions, these problems are generally eliminated. After women have been in units for a period of time, it is usually found that the efficiency of the unit does not suffer. In fact, women are often able to find alternative ways to do heavy work that benefit both men and women. Many units have found discipline is easier when women become part of the group, though infractions committed by women may differ from those committed by men. Women who join the service may be more oriented toward discipline than their younger male colleagues, or it may be that the women, sensing that they are being tested, behave more circumspectly than they might otherwise. In any event, the result for the service is easily disciplined troops accomplishing the missions they are charged with, and, as time passes, attitudes of men in the military usually begin to change.

Suggestions: The persons in the Navy who often have to make day-to-day decisions are the petty officers. It is almost a cliché to say the Navy is run by its chiefs. They are the people that will make the "on the spot" decisions that will affect women, as women come more and more into the mainstream of military life. It is important that this mid-level staff be taught to deal with women as subordinates, peers, *and* as superiors. They must adjust to the idea that women are a *permanent,* valuable part of the military establishment, and that the services will continue to support equality of opportunity for women. A model for this program, while by no means perfect, is available in the race relations programs that the services have instituted. Noncommissioned officers who must deal with the problems of women who work for them must be made fully aware of military rules governing women, e.g., regulations concerning uniforms and pregnancy leaves.

Enabling women to advance in responsibility and rank, if qualified, even though lacking traditional flying or shipboard experience, would help alleviate the dissatisfaction many servicewomen feel when they perceive a lack of opportunity. It would also provide visible role models for younger women and help eliminate difficulties many junior men have in dealing with women as fellow service members. Finally, providing greater promotional opportunities for women would be a concrete example of the military's good faith in instigating opportunities for women. A caveat is necessary at this point, however. Promotion of women no matter how qualified, if lacking traditional career patterns, is sure to antagonize men who are perhaps less well qualified, but have followed the traditional paths of air/sea duty.

If men feel that women are getting something they are not receiving, even after they have performed all the activities necessary to attain that goal, animosities may build up. A feasible solution to this problem would be the removal of all restrictions on the duties women in the services could perform. This would, of course, necessitate research into the strength and skills necessary to perform military jobs, and perhaps the development of programs to teach skills that women lack. Basic mechanical knowledge is often used as a measure of technical skill for determining placement in a training school, although mechanical knowledge is often only peripherally, if at all, related to the skills taught at the school. Other tests, not biased against women, that better predict skills necessary for a job must be found. In addition, if mechanical knowledge or other skills are found to be extremely important in the performance of military jobs, remedial courses must be made available to women in boot camp to provide them with skills that men have already achieved in high school shop classes, and attendance of these classes must be without demerit on the women's records. Such training could be handled similarly to the remedial language arts training now offered in boot camp to enlistees that it would benefit. If the military is really interested in acquiring a certain type of female enlistee, recruiting drives must be designed to encourage enlistment by women of that particular caliber. Enlistment without adequate information about the services, followed by disillusionment and/or discharge, helps neither the services nor the enlistee.

Many Questions, Few Answers: How can the military maintain two opposing policies (no women in combat, no discrimination against women) and still maintain an effective fighting force? Fitting women into nontraditional roles effectively means establishing task requirements with respect to strength, stamina, muscular coordination, height, weight, and energy. What is the normal range of female abilities (i.e., mechanical ability, spatial visualization, manual dexterity, etc.)? In past years the norms have usually been based on male populations only. Do women demonstrate the same decrement following sleep deprivation that men do? Are there perhaps times or circumstances when women can maintain sustained attention or performance levels which *exceed* those of their male counterparts? How do women perform under stress or in stressful environments? What positive steps might be taken to achieve attitudinal changes among men and women to deal better with the new roles of women? What are the factors which account for the non-completion of women's enlistment contracts? We have studied military father absence extensively, but we have little if any knowledge about the health and functionality of the all-military family (e.g., the family in which *both* spouses are service personnel) or the family where the husband is the military "dependent."

Adequate, realistic, up-to-date information about women's abilities and opportunities for women within the military must be made available to recruiters and, through the media, to potential enlistees. Needless to say, research into the characteristics and capabilities necessary to perform a job effectively is expensive. So, too, is the underutilization of half of one's possible work force. Thus, more than a mere plea for women's rights, what is now needed are some well-designed research efforts to answer the questions being asked, which, in turn, may solve, or at least alleviate, the problems of woman/job and woman/machine mismatch within the military organization.

Dr. Hunter is a clinical research psychologist who has been affiliated with the Naval Health Research Center at San Diego since 1967. She received her undergraduate training at the University of California, Berkeley, master of science degree from San Diego State University, and doctoral degree from United States International University. She has coedited two books on the military family and a number of journal articles in the areas of family research, psychophysiology, learning disabilities, and social isolation. Dr. Hunter devoted five years' co-research in the areas of the psychophysiology of sleep and the physiological aspects of developmental dyslexia prior to her assignment to the Center for Prisoner of War Studies. She now serves as both the Head of Family Studies and as the Assistant Director for Administration of the center.

Ms. Million is a third-year law student at the University of San Diego. She has an A.B. from San Diego State University in political science/sociology. She has been at the Center for Prisoner of War Studies, Naval Health Research Center, San Diego since 1972. First, she was a medical data analyst with duties involving the compilation, coding, and analysis of medical, social, and psychological material for a five-year study of the Vietnam POW. More recently, she has been a member of the Family Studies Branch, involved in research, writing, and editing of a number of reports on the military family. Ms. Million has also been a case worker for the "Women's Legal Center of San Diego" and a law clerk in the San Diego County District Attorney's Office, Consumer Fraud Division.

"Just Call Her Captain"

9

Captain Nori Ann Reed, USN

U.S. Naval Institute *Proceedings*
(December 2007): 28–31

Today it's common to see women on board ships of all types, including combatants. But it wasn't always that way. One of the first generation of women officers to go to sea looks back at how far they've come.

I CAN STILL HEAR THE CO's words clearly from the day I reported to my first ship in August 1980. It was one of the first things he said to me as I sat in my blues, perched with perfect posture on the chair in his in-port cabin during my in call.

"I don't think the Women at Sea program will be around very long."

I remember blinking and being unsure of what to say. Hmmm, this did not sound good; maybe he knew something I didn't. I had been commissioned in May 1978, and women hadn't been assigned to ships then, so I had gone to shore duty on Guam. In October 1978, a few women were assigned to a limited number of ships under the Women at Sea program. I was thrilled when I was accepted into the program and received my orders to a sub tender, one of the few classes of ship then available for the embarkation of women. Now, I wondered, was it all going to abruptly end?

Listening to the CO's comments, I knew he didn't say them maliciously. Instead, I recognized it as his sincere attempt to prepare me for what he believed would have to be another career path. Knowing that still didn't make it sound any better. He went on to say that, while he didn't think the program would be around much longer, he would treat me as he had every other junior officer he had commanded. That meant while I was on his ship I would be expected to qualify as engineering officer of the watch, officer of the deck under way, and, if around long enough, surface warfare officer. And because of his strong belief in his responsibilities toward a junior officer's qualification, if not his belief in women qualifying as surface warfare officers, most everyone on the ship followed his lead. In fact, some of my shipmates were downright enthusiastic about the five women officers being assigned there.

Two of the most enthusiastic were my department head and his assistant. They were determined to give both Ensign (now-Captain) Ann O'Connor and me every possible opportunity to get under way on other ships since ours did so infrequently. As the Repair Department team, they were not above bribing other ships to take us to sea. While most people are uncertain as to their value, Ann and I could take comfort in knowing exactly how much we were worth—27 brass ship's seal plaques made by our tender's foundry for the lucky destroyer that took us out to sea for six weeks so we could perform shiphandling skills often not feasible on our own ship.

But those words from that first CO would haunt me. Sometimes they seemed to be confirmed. At one point, a later skipper of that same ship decided that when the Soviet Union attacked (and he was certain it would), all the women officers (we had no women enlisted) would need to be put ashore. While it seemed very odd that the Navy would assign women officers (in our case a physician, damage control assistant, main propulsion assistant, navigator, automated data processing officer, and electrical repair officer) to a ship, then before the ship fought its way through the Greenland-Iceland-United Kingdom gap, would have them all taken off, we thought maybe that was what the Navy planned to do with us. (At one point, I remembered hoping that we would be ashore when the balloon went up because there was no doubt in my mind if we were at sea, this CO would load us in a lifeboat and give us a magnetic compass and directions to the closest land. At least we had the navigator!) While it would eventually be settled in my CO's mind that we shouldn't be thrown off in time of war, it didn't make me feel any more secure in my choice of professions.

At the same time, those words would spur me to qualify in every watch and qualification I could, as quickly as possible, and at every opportunity. Who knew when it all might be taken away? As a result, I proudly earned my SWO pin on 16 September 1981. I wear that same pin today, having worn it through countless sea details and hundreds of underway replenishments. It may not be pretty, but except for my Command-at-Sea pin, I value it more than any other decoration or badge I wear.

Great Strides

Certainly things have changed since that day in 1980. No longer must destroyers be bribed to take women to sea. They are already there. Often they command the ship. However, knowing something has changed is not the same as believing in the change. I don't think I ever really believed that women would be allowed to remain at sea until just before I took command for the first time. It actually didn't having anything to do with taking command. Instead, this epiphany came from a rather small, innocuous event.

In preparation for taking command of the USS *Kiska* (AE-35) in the spring of 1995, I was attending steam engineering hot plant training in Great Lakes on the day shift and entered the small storage closet that served double duty as the women's dressing room. Inside was an ensign changing into coveralls to take the evening shift. After we exchanged pleasantries, I asked where she was going and she replied she was going to be the boilers officer on the USS *Boxer* (LHD-4). I said I was so happy to hear that and that I wished I had been able to be the boilers officer on the *Boxer* as an ensign. She gave me that look only junior officers can give their superiors, indicating a moment of impossible stupidity, and after an instant, asked:

"Well, um, ma'am, um . . . why didn't you?"

I laughed so hard I nearly fell over and tried to figure out how I would even begin to answer that question. I finally stammered something to the effect that going to any ship hadn't been possible when I was an ensign and left it that. As the young officer made her escape from the closet with a polite, disbelieving smile, it struck me that things really had changed. Here was a generation of officers, both men and women, who would never know a time when women hadn't been stationed on ships. It was simply a part of their normal life on ships of all classes. And I was so thankful that this ensign would never feel that same sense of concern that her profession would suddenly be taken away from her.

Make no mistake, the question of women belonging still lurks out there. Sometimes it is in the smallest, most innocent of things. One example is the case studies book used in the leadership and ethics course taught at the Naval Academy and NROTC. In it is a case study discussion on whether or not a female Judge Advocate General officer should be used on a mission in the Middle East. In one respect, this is a reasonable question since not every partner nation shares our values. Having been the Commander of the Naval Logistics Forces in Central Command for nearly two years from August 2003 until June 2005 and the captain of the USS *Detroit* (AOE-4) deployed to the Middle East after 9/11, I understand there are significant cultural differences between the United States and other countries. There have been businessmen who could not shake my hand or look me in the eye. I have had harbor pilots come on board my ship and address my male third class yeoman phone talker rather than speak to me or my female conning officer. There was no point in taking offense. It was, after all, their culture I was entering and I was respectful of it. The businessman would either agree or disagree to provide services or products, and most were quite willing to overlook my gender to gain the contract. In the case of the pilot, I didn't care to whom he spoke. I could still hear him, and I was absolutely certain it was my voice that my crew would ultimately listen for and follow.

A Diverse Navy

In addition to task force commanders and commanding officers, women have served in numerous leadership roles in the Middle East including amphibious squadron commanders. Since women are already there, an even more reasonable question is how we deal with differences of culture, not in regard to women or any specific group of naval personnel, but to values held and customs in general.

As we move toward a more diverse Navy, reflective of our great nation, we work not only to recruit, but more so to retain that diverse population. And to achieve this it is important for our people to believe in our belief in diversity. To that end, words are significant, especially if you are on the receiving end of them. It took years for me to get over that simple sentence by my first ship CO. But seeing is believing. If you don't see it, you don't necessarily dream of achieving it. Just this past summer, as I was conducting NROTC midshipmen training in San Diego, I had junior women, both officer and enlisted, coming up to me while I was in uniform in a variety of locations from on board ship to the Exchange, asking what I did. As I conducted mini-mentoring sessions, I received reactions such as "I didn't know we could do that" and even "I have never seen a female captain." I found that last statement very hard to believe.

"You mean a female captain who is a surface warfare officer, right?" I would ask.

"No, ma'am, any female captain. I might have seen a female captain once, but I'm not sure."

The Navy is working hard to change those reactions. But it can't just be an institutional thing. It has to be something that each of us, whether a member of the majority or minority, embraces. Watching our words, yes, but also honestly examining our own beliefs to see where we can make a difference. I am no different. For many years, I would cringe if asked to participate in anything that remotely smacked of publicizing a "female SWO." I just wanted to be known as a SWO and I would get on my high horse, and ask "why we needed to be labeled," or say "leave me alone to do my job." It was never a successful speech with my seniors but at least all of them were polite when they listened to it. Now I know there is more to my responsibility than just being a SWO of any label. Certainly my greatest responsibility is to be the best CO possible to the men and women under my command. But I have come to realize and cherish that, as the commander of an NROTC unit, I have a part in educating both the Navy and the civilian community here in New Mexico. That is my part, but each of us needs to determine how we best fit into the puzzle.

Follow Your Dream

Despite the sometimes chaotic beginning, I certainly can trace part of my success to that first ship. First and foremost, I found the career I loved. While I enjoyed my first tour on Guam, it could in no way match my time at sea. The responsibility, the camaraderie, the sense of doing something worthwhile all made for a career that was beyond my wildest imagination. I especially loved being a logistician. Sometimes I would look longingly at that sexy destroyer 140–180 feet away from me, connected up by rigs and wonder "what if." But being a "loggie" meant you didn't practice your profession, you performed it, every single day. There is something incredibly satisfying about racing across the Indian Ocean, finding the carrier in the creeping moments of the dawn, sliding through the rendezvous point and closing up Romeo precisely on time, succeeding in spite of the carrier's best efforts to thwart your hard work by changing the rendezvous and course three times in the last hour. Serving in the "blue-collar Navy" with such individuals as steam engineers and boatswain's mates is an honor I will always treasure.

Yes, that first CO's words were in some respects discouraging, but his actions belied those words. Before the first women walked on board his ship, he had stressed to the crew the respect we should be shown and explained that we were to get the appropriate opportunities to succeed. He might have had personal reservations, but he ensured his professional obligations came first. Those opportunities each of the original four female SWOs on the ship were given must have worked—three of the four would remain in the Navy, have surface command, and go on to selection for major command. It taught me to give people their own opportunities to follow their dreams, even if I am uncertain of their ultimate success.

I have thought of that first CO each time I have taken command. He gave me my start and in many ways pushed me to go farther than I ever thought possible. But I am really, truly glad he was wrong!

Captain Reed served on six ships, with the honor of commanding three of them: the USS *Kiska*, USS *Willamette* (AO-180), and USS *Detroit* (AOE-4). She also was the Commander Logistics Forces, United States Naval Forces Central Command (CTF 53) and is the commander of the University of New Mexico's NROTC unit.

"Women Are Fitting in Fine"

10

An Interview with Vice Admiral
Ann Rondeau, U.S. Navy

Fred Schultz

U.S. Naval Institute *Proceedings*
(December 2007): 22–26

*In several circles, she's the odds-on favorite to be the Defense
Department's first woman four-star flag officer. She is the former
Commander of Naval Training Center Great Lakes and was
the first Commander of the Naval Service Training Command.
She was also Commander of the Naval Personnel Development
Command and served as the Director of Navy Staff. Vice Admi-
ral Ann Rondeau is currently Deputy Commander, U.S. Trans-
portation Command, which manages global air, land, and sea
transportation for DOD. She met recently with* Proceedings
*Senior Editor Fred Schultz at Andrews Air Force Base outside
Washington before a flight back to her headquarters at Scott Air
Force Base, Illinois.*

PROCEEDINGS: HOW DID someone who was born in San Antonio,
grew up on the Hudson River, and graduated from Eisenhower College
in New York's Finger Lakes region end up making the Navy a career?

VADM Rondeau: My father had been in World War II, and both my
father and mother were civil servants. So the whole notion of public

service was attractive. I was also fascinated by American history, even as a little girl. And any history of almost any nation includes military history.

My parents took us to West Point for Sunday summer concerts, and my father took me to the museum there. I sang "The Messiah" with the Corps of Cadets for two years. They had no women back then, so they brought in girls and women from high schools and colleges who could sing. West Point was very familiar to me, and as a result, the military environment was not uncomfortable for me.

I went to Eisenhower College toward the end of the Vietnam War, when the draft was ending and nobody was joining the military. Well, I did. A classmate friend of mine joined the Coast Guard, and he and I were the only two from our class who joined. I had been accepted to go to graduate school, but I didn't have any money. So I began asking myself what I should do. My sister was on a full Navy scholarship for nursing out of Hartwick College, and she suggested I take a look at the military.

I interviewed for another service and did horribly. I did not interview well, and I tested very badly. I then went to the Navy office and took the test, which was more about aptitude and actually tested my strengths. I did very well.

Before this I had done job interviews from college, and everybody wanted to start me out at a low clerical level across the board, in what at that time were traditional roles for women. The Navy said it would bring me in as an officer and treat me as an officer. This gave me a sense that the Navy wanted me and that I would be part of the larger culture.

Proceedings: As a woman, what were your impressions of Officer Candidate School [OCS]?

VADM Rondeau: The only avenue for me was OCS. West Point and the Naval Academy were both closed to women. And I could not even have gone in through the ROTC program. Had I been allowed to go to into an academy, I probably would have applied to West Point. I had read about the great West Pointers MacArthur and Eisenhower and Bradley, and that appealed to me.

Back then, OCS expanded or contracted depending on what the end-strength requirements were, and it still does to some extent. It's also a fast way to make officers. You can do it in five months versus four years. When I came in, this is where minorities and women went to be trained for a commission. I was in the second co-ed OCS class.

Proceedings: How have women's roles evolved since you were commissioned in 1974?

VADM Rondeau: Women's roles—and men's roles, for that matter—have changed constantly over time. At the time I was commissioned, boys never went to home economics class to learn how to cook. Now they do. Roles evolve depending on the culture or because of economic or social changes. As I said before, I grew up during the Vietnam era, but I also came of age during the Civil Rights movement, a time of tumultuous changes in this country.

The feminist movement was becoming more aggressive. But when I first came into the Navy, the traditional roles for women were still in place—administrative support, supply, logistics, personnel. My first job, however, was on a four-star staff as the communications security officer. I filled a Fleet lieutenant's job. All of a sudden, I was put into a quasi-Fleet-support role. That in itself was an indication of change.

Also at this time, women officers could not wear khakis. We had to wear skirts, and we had to wear gloves. If you were pregnant, you were forced to leave. This became a big issue after Vietnam, when the Navy was losing so many people. Not only was it cutting personnel, it was losing them, too. So the Navy was forced to find ways to retain its people.

Women have had other biases to contend with. Before Title X was finally in place, women were not allowed to go to sea or fill combat roles. So it was really a struggle for women to become operational. I remember not receiving the Navy Achievement Medal. My bosses had put me in for it, and I earned it. But a four-star admiral wrote on the top: "No officer who's never been to sea should ever get a personal award." I've never forgotten that. It was telling. I understood that he had a view deeply embedded in his experience. What did that mean to me? That meant I had to also think about my own experience and apply it broadly.

Yes, I was disappointed. But at that moment I decided I had to stretch myself out of my current experience and seek to understand things intellectually and emotionally through empathy and through observation. I decided I needed to immerse myself in the environment around me to understand the people and know what was going on. I began to do things unconventionally.

I went on board ships tied to the pier at Pearl Harbor and sought out peers to ask questions. "Tell me about what you do as a combat systems officer. Tell me what you do as a gunnery officer. Show me the bridge, and show me what happens there." By getting out from where I was and going to where I needed to be, I began making a name for myself on the waterfront and started to develop a reputation.

I earned my surface warfare pin because of the goodness of a Fleet commodore. I told him I wanted to learn about the Navy and about ships at sea. At the time, the Bureau of Navy Personnel had a provision that if you were assigned to a sea-going unit or assigned to an operational unit, you could get your surface warfare pin.

I was doing things that women were not allowed to do, like going on deployments with a VP [patrol] squadron. I was not wearing aviation wings, but I was assigned to the squadron as an air intelligence officer [AIO]. We were seeing a tremendous exodus of aviators—pilots, in particular—to the airline industry.

I was reliable, kind of a special-teams player as well as AIO, and I had done flight schedules and extra squadron duty when the aviators were up flying. One day, the skipper said, "Okay we're going to make you the operations officer." I was probably the only ops officer in naval aviation not wearing wings. But he needed somebody to be the acting ops officer on deployment. That was not about being a man or woman. He said he had aviators to fly the planes. He needed an ops officer, and I was it, which was extraordinary for the time.

Back then in the civilian world, few women were going to law school or medical school. But some women decided to test the system, and that's what makes America great. I wasn't denied very much because I was a woman, even though some policies in place were, shall we say, less than accommodating in the early days.

Proceedings: Like what?

VADM Rondeau: The Navy did not have an ethos or culture that valued women. For one, the uniforms were odd. They were not functional for women officers who wanted to do operational work. And we had no women's locker rooms, gyms, or fitness centers. I guess no one thought women worked out. The Navy is a culture that values sports, but we didn't have women's sports or facilities for them in the early days. Some of the really creative commands made signs that read, "Women's Locker Room" or "Men's Locker Room" and switched them on the door, depending on who was inside.

Proceedings: How did you cope with that lack of accommodation?

VADM Rondeau: I think it was mostly about attitude. Can women be operational? Can women make tough decisions? At first women didn't always do very well with attitude. We did great work but we weren't doing professional things like reading about naval tactics, about maritime and military history. So we could not have a professional conversation about ships going to sea. We did not have the vocabulary, the lexicon, or the roots of our profession that the culture required for acceptance.

I'm not arguing that women should become men and men become women. But back then for women to succeed they needed to be adaptable. Now, young women are joining the Navy in an age when they have girls' sports or women's sports in high schools and colleges. They also have such women role models as lawyers and politicians and doctors and judges.

Think about the women back then who came into the Navy just to serve, with no delusions. At one time, women Navy officers had no expectation of becoming a captain. When I joined, we had something like two woman Navy captains, not counting nurses. They also had no expectations of having command. They just came in to serve and they adapted. They fit in with the least amount of training, and they performed brilliantly.

We did not send the female officers to schools where they could learn their trade or their craft. I went to very few schools as a young officer. It's a different story now for our young women who are so fully integrated and assimilated. I have complete admiration for the women who came before me who had few indications, no assessment, no training, and little acculturation to succeed in this very competitive, performance-driven environment.

Proceedings: You said earlier that you had developed a reputation for being on the edge. Some of the women officers today seem a bit stand-offish about it, saying, "Why do we want to draw attention to ourselves?"

VADM Rondeau: There is stand-offishness, and I have it as well. It took a while, but at a certain point, women decided that they just wanted to fit in. [Former Chief of Naval Operations] Admiral Thomas Hayward once said something along the lines of, "we'll know that women will have made it in the Navy when we're no longer talking about them as women and we're talking about them as naval officers or as Sailors." So that's what you're hearing from women today. It's like any team sport. You don't want to convey a sense that you don't belong to the team. When I said I put myself on the edge, I should have also said that I had great bosses who let me do that. In the larger context of our Navy, I was not shut down.

What you hear from women is all about how they contribute to the team, about how they measure up against their teammates, and about how they add value to the team effort. Good leaders would rather have you not make a big deal about them but rather make a big deal about what was done. All of us want to fit in, and we want to be told that we were a good naval officer, not a good female naval officer.

Proceedings: Former Chief of Naval Operations Admiral Vern Clark said a few years ago that the issue of women in the Navy is no longer an issue. Is that true?

VADM Rondeau: Yes, it is. With Title IX changing sports and Title X's changes, access is no longer an issue. Now, it is all about performance. We now say we value women in the Navy. And we value families. We're now talking about parental leave and not just about maternity leave. I'll never forget an article I read in the mid 1990s in *Navy Times*. In an interview, the admiral in charge of Air Forces Atlantic said something phenomenal was happening in naval aviation that he had never seen before. He was forced to deal with a wave of young male junior officers leaving naval aviation because they wanted to be good fathers, and they couldn't be good parents and be away from home for so long. He said we needed to begin to address this issue.

Getting back to women, being able to fly planes and drive ships is also no longer an issue. We still face some issues now being discussed, like women in submarines and various other roles. But in terms of performance and assimilation, women are fitting in fine.

Proceedings: What about retention?

VADM Rondeau: We still have a problem with how we retain our best talent, male and female. The junior officer of 2007 is very different from the JO of 1987. Frankly, we now have a junior officer corps that has more operational experience, in real operations, than the average senior officer. Even though women's health issues and pregnancy issues still exist, it's no longer about women in the Navy, per se. It's much larger than that. It's what our cadre of young officers says to us. They want education, they want some stability, they want some parental leave time, and they want some of the same flexibility found in the civilian sector. This is part of the conversation about active and reserve. Maybe we should allow JOs to go back and forth between active duty and reserve, rather than taking sabbaticals where they lose seniority. This could allow for accommodations to family or time for that master's degree.

Proceedings: Have enough accommodations been made for women?

VADM Rondeau: For the most part, yes. But nothing ever just disappears. The issues have been addressed over time. When I came into the Navy, it was a drug-infested, alcohol-infested service. And post-Vietnam depression was an issue. We had plane crashes at the time because of wear-and-tear on aircraft. We had a tired force from Vietnam, and we had a force that had racial problems. In some ways, the Navy was on its knees. But the service pressed ahead, and great leaders came to the fore. I think one of the most unheralded CNOs was Admiral Jim Holloway, because he was the leader who helped get the Navy healthy again. Few problems go away quickly. You have to work at it. If you begin to establish a dialog about race or drugs or alcohol or sexual harassment, other dialog occurs along the way. Let's face it. We've been through a crucible of experiences. We went through Tailhook and we went through the explosion on the battleship *Iowa*, experiences that fundamentally altered our Navy. And we got better.

I am so impressed that we've had leaders—male and female—who have brought our Navy along to a much healthier place. I would put their effort up against any effort by any civilian profession.

Proceedings: What advice would you give to a young woman considering the Navy as a career?

VADM Rondeau: I think being able to laugh is really important. Just laughing and being able to laugh at yourself. We women tend to be pretty intense at times. I once had a guy tell me, "Just lighten up, Ann, and you'll do fine." I've never forgotten that. The young ladies today are so talented. I'd tell them to believe in yourselves, have a good sense of humor, pitch in, and be good to yourselves.

Proceedings: Who were your role models?

VADM Rondeau: [Former U.S. Naval Institute Board member] Captain Betsy Wiley, for one. I was an ensign and I knew all about Betsy Wiley, a woman pioneer in the Navy. My recruiter was also fabulous. She was Carolina Claire, who happened to be Betsy Wiley's sister-in-law. She was a good-looking and vibrant naval officer. The other name that comes to mind as a role model for women is Rear Admiral Bobbie Hazard, who was universally respected by men. Here was an old-fashioned kind of naval officer who had adapted. She was dignified and smart, too. She had a presence and a sense of herself.

I had a lot of male role models, too. Some were mentors, some were tormentors [laughter]. One of the mentors was Rear Admiral Ron Marryott, a fabulous guy. When I was a one-star, Admiral Archie Clemins was also a great mentor. He put me through my paces and made sure I knew my stuff. Rear Admiral Sam Packer did the same thing, as did Rear Admiral Charles Prindle and Captain Ed Gibson. Captain John Grotenhuis was truly the leader who made a huge difference in my professional life. There were so many male peers and other great leaders. And there were always the chiefs and master chiefs, too many to name. I was very lucky.

Proceedings: Is the Navy ready for a woman CNO?

VADM Rondeau: Culturally, the Navy is ready for a woman CNO, I think. We're not yet there in terms of someone who's qualified for the job. We need someone who has gone through the extra wickets. We demand that of a man. We need to know what he did and what he didn't do. We make those distinctions all the time. I think it's going to have to be somebody who's proven herself and done it double. There will be a female CNO, but she must have had the right operational command—a battle group, perhaps.

Proceedings: Is command at sea that important?

VADM Rondeau: Absolutely. That's important for male CNOs. And it's just as critical for the first female CNO or for the first female Vice Chief. CNO candidates must have built reputations. Are they respected? What did they do to gain that respect? We don't have one career path that makes you the Chief of Naval Operations. For example, we won't see a female submariner CNO for a long time. Do I think we could see a female naval aviator or surface warrior as CNO? Absolutely.

11 "Epilogue"

Susan H. Godson

(Selection from *Serving Proudly:
A History of Women in the U.S. Navy,*
Naval Institute Press, 2001): 279–91

THE RECENT PAST presents difficulties for any historian. Vital source materials are not yet available, and insufficient time has passed to place events in historical perspective. Nevertheless, the 1990s, chock-full of fast-moving actions, saw old barriers crumble and new opportunities open. At the same time, long-standing problems persisted or intensified. A brief summary of major events from 1990 to 1995 can indicate roads traveled by Navy women and perhaps offer a glimpse of possibilities yet to come.

Navy Women

With the collapse of the Soviet Union in the late 1980s, the United States had no major cold war adversary, and the military buildup slowed. All the services began downsizing, and the Navy reduced its active ships from 570 in 1990 to 392 in 1995. Concurrently, personnel requirements dropped, and the 604,562 on active duty fell to 444,661, which included 55,548 women, or 12.5 percent of total strength. Although the number of women decreased from 60,411, their percentage of the active-duty force rose.

Desert Storm

The defining test for Navy, as well as all military, women came in 1990–91 with American participation in the Persian Gulf War. Ensuring the flow of Middle Eastern oil to the West had long been a keystone of American policy, and that flow was jeopardized in August 1990, when President Saddam Hussein sent the Iraqi army to invade the tiny emirate Kuwait. When neighboring Saudi Arabia requested American help, President George Bush dispatched a massive military force as part of a thirty-three-nation coalition to drive Saddam from Kuwait. Vital components were the 167 ships and more than 75,000 naval personnel deployed to the area.

Eventually, more than 1 million military men and women took part in the Persian Gulf War, including more than 37,000 American women. Of naval personnel, women accounted for about 3,700. This largest deployment of women in history included, for the first time, military mothers. The war revealed how completely women had become integrated into the armed forces and how essential their services were. Military women suffered casualties: thirteen Army women died; two were captured by the Iraqis.

Navy women filled a variety of roles in the combat area. Hundreds served in support ships—ammunition, supply, tenders, and oilers—and in Military Sealift and Combat Logistics Force vessels. Others were in two helicopter combat support squadrons, in Construction Battalions at Al Jubayl, and in a cargo-handling and port unit at Bahrain.

For the Medical Department, the operational readiness concept, a part of planning and training for years, worked efficiently. About 250 nurses, both females and males, served in the hospital ships *Comfort* (T-AH 20) and *Mercy* (T-AH 19), dispatched to the gulf in September 1990. These state-of-the-art one-thousand-bed floating hospitals were prepared to care for massive casualties that, fortunately, never materialized. Twenty-one male nurses staffed amphibious assault ships *Guam* (LPH 9) and *Iwo Jima* (LPH 2). The First Marine Amphibious Brigade and the First and Second Force Service Support Groups utilized thirty-one additional male nurses.

Far busier were the three fleet hospitals erected in the region. Fleet Hospital Five, composed largely of active-duty personnel, including 152 nurses, reached Saudi Arabia in August 1990, and Fleet Hospitals Six and Fifteen, staffed mainly by reservists, arrived a few months later. Designed to handle battle casualties, these hospitals instead dealt primarily with illnesses or injuries not caused by combat. Navy nurses played vital roles in these medical facilities, all located within range of Iraqi Scud missiles. Reserve nurses became critically important, recalled Nurse Corps director Rear Adm. Mary F. Hall, to staff fleet hospitals and to backfill in stateside hospitals for nurses deployed to the gulf.

For all Navy women and nurses, the Gulf War demonstrated women's capabilities in a combat zone and exposed them to enemy fire. "We could not have won without them," said Secretary of Defense Richard Cheney. Their service, along with that of Army, Air Force, and Marine Corps women, brought to the forefront that old prohibition, combat exclusion.

Combat Exclusion Repealed

Women's superior performance during the war elicited immediate demands to reexamine the outmoded 1948 combat exclusion law and triggered vigorous debate. In April 1991, DACOWITS pushed for an end to the restrictions. By the end of the year, at the urging of Rep. Patricia Schroeder (D-Colo.) and others, both houses of Congress had debated the issue, authorized women to fly in combat planes, and mandated a presidential commission to study women's roles in the military.

In November 1992, after eight months of acrimonious debate, the Presidential Commission on the Assignment of Women in the Armed Forces issued its report. The document recommended reenacting section 8549 of Title 10, barring women from combatant aircraft, but suggested allowing women to serve in all combatant ships except submarines and amphibious vessels. It also recommended exclusion of women from direct land combat.

Following this contradictory and divisive report, Secretary of Defense Les Aspin took matters into his own hands and in April 1993 ordered the services to assign women to combat air squadrons. The services, he declared, must also offer more specialties to women, and the Navy must open additional ships to them. In November Congress repealed section 6015 of Title 10 and thereby opened combatant ships to women. Aspin went further still in January 1994 when he announced that the Defense Department was rescinding the risk rule of 1988 and instituting a new policy that barred women only from direct ground combat. There was opposition from service women as well as men.

It had taken since 1948 to remove most combat restrictions for women. The repeal of section 6015 and changes in Defense Department policy put women on more equal footing with men for career and promotion opportunities. By 1995, Navy women could serve in twenty-four of the twenty-six officer communities and in ninety-one of the ninety-four enlisted ratings. It is far too soon to assess the long-term impact of these sweeping changes, but at least the full assimilation of women into the Navy and the other services moved forward, although military women continued to fill only a low number of combat-related jobs.

Sea Duty

Nowhere was progress more apparent than with women on ships. In 1990, 331 officers and 7,803 enlisted served on 117 ships. Although downsizing the Navy during the early 1990s reduced the number of billets available to women, the 1993 repeal of the combat exclusion law opened a whole array of opportunities, most notably permanent assignment in combatant ships. The *Dwight D. Eisenhower* (CVN 69) was the first combatant with a mixed crew and air wing, and in October 1994 more than four hundred women sailed on the aircraft carrier on a six-month deployment to the Caribbean, the Middle East, and the Adriatic. "I've waited my whole career to go on a carrier," exclaimed CPO Chris Jackson. The *Eisenhower* successfully completed her mission with no diminution of readiness or performance. There was some resistance from male crew, recalled CP02 Sandra Plunkett, who was told, "This is our world. This is a man's world." But the commanding officer, Capt. Alan M. Gemmill, called the deployment a success. "I think we've become a little more civilized," he added.

As more combatants were reconfigured to accommodate female crew members, more women were assigned permanently to these warships. By the end of 1995, 374 officers and 2,332 enlisted were serving on 40 combat ships. An additional 218 officers and 5,425 enlisted were on 72 noncombatants. Altogether, 8,349 women were serving on 112 naval vessels. The Navy wisely insisted on assigning sufficient female officers and enlisted to ships to prevent women's feeling isolated. The only combat ships that remained closed to them were submarines and mine warfare vessels.

Duty afloat, especially in combatants, was important to women, not only because of equal opportunity but also because it made them more competitive in promotions. "It was vital to have the same experiences as men," recalled Senior Chief PO Mary B. Prise, who served on four ships.

In Aviation

The crumbling of the combat exclusion law had an equally positive effect on women in naval aviation. In 1990, there had been 173 qualified female pilots and 80 naval flight officers, with more in the training pipeline, as well as 7,733 enlisted in aviation ratings. Two years after Congress repealed the ban on female pilots for all the services in 1991, the Navy began assigning women aviators to combat aircraft squadrons.

Aviators such as Lt. Sharron Workman, the first to qualify for carrier operations, was one of six female pilots assigned to the *Eisenhower* in the fall of 1994. While deployed in 1994–95, female fliers from the ship took part in 156 of the 2,856 sorties over Iraq and Bosnia. Another pioneering combat pilot was Lt. Kara S. Hultgreen, the first woman F-14 pilot to qualify for combat duty. Her plane crashed during a training flight from the carrier *Abraham Lincoln* (CVN 72) in the Pacific in October 1994. Although later investigations unearthed mechanical problems as the cause of the crash, other aviators started a whisper campaign discrediting her qualifications, raising anew questions about women in combat aviation.

Lt. (jg) Carey Lohrenz, another F-14 pilot and close friend of Hultgreen assigned to Abraham Lincoln, later told of being ostracized and derided, a fate the three remaining female pilots shared. Never a part of the squadron "jockdam," all these women eventually left the carrier. Reinforcing such rejection was a naval helicopter commander's refusal to fly with women on combat missions during the American deployment ashore in Haiti in the summer of 1994. He cited his religious belief that men are to protect women.

In spite of lack of acceptance by their male peers, the number of women in aviation continued to increase. By 1995, there were 206 pilots, including 82 combat aviators, 77 naval flight officers, and 9,502 enlisted.

Attesting to the competence of naval women, NASA selected three for its space shuttle program. In 1992, Lt. Cdr. Wendy B. Lawrence, a CH-46/SH-2 pilot, started training as an astronaut. Three years later, Lt. Cdr. Susan L. Still, an F-14 pilot, and Cdr. Kathryn P. Hire began preparations to become astronauts.

Naval Academy

The U.S. Naval Academy, that bellwether of women's progress and acceptance, admitted even more budding female officers. In spite of a number of incidents that put the Academy on the defensive, women continued to apply for admission, especially after the 1993 repeal of the combat exclusion law. By 1995, there were 571 women at the Academy, 14 percent of the brigade. Still blocking women's true acceptance was the resentment festering even after twenty years of women's attendance.

Navy Nurse Corps

After participating in the Persian Gulf War, the Nurse Corps returned to its peacetime activities. There had been 3,102 active duty nurses, including 790 men, in 1990. Five years later, there were 3,319. The corps had ample nurses, which was a marked change from the shortage in the late 1980s. In fact, recruiting goals for 1994 included only 20 nursing specialists. The Navy soon instituted a reduction of strength that would gradually cost the corps 199 billets.

Nurse Corps Leaders

Replacing Rear Adm. Mary F. Hall was Rear Adm. Mariann Stratton, who became Nurse Corps director and deputy commander for personnel management in September 1991. Born in 1945, Stratton, a Houston, Texas, native, entered naval service in 1964 with a Navy Nurse Corps Candidate Scholarship and graduated from Sacred Heart Dominican College with nursing and English degrees. She later earned master's degrees in human resource management and nursing. Beginning active duty in 1966, Stratton served in a variety of hospitals overseas, including Japan, Ethiopia, Greece, and Italy, and stateside. Her last assignment was director of nursing services, Naval Hospital, San Diego.

Although the 1990s saw downsizing and budget cuts, Stratton advocated a philosophy of "total quality leadership" and encouraged her nurses to devise innovative means of advancing patient care such as nursing case management programs. She emphasized research and publishing to spread innovations in nursing. Reflecting her vision, in 1993 she devised *Nurse Corps Strategic Plan—Charting New Horizons,* an organization plan and course of action.

Following Stratton came Rear Adm. Joan Marie Engel, who became director of the Navy Nurse Corps in September 1994. The following year, the Health Sciences, Education and Training Command was disestablished and with it Engel's collateral job of deputy commander for personnel management. In its place came the billet of assistant chief for Education, Training, and Personnel.

Born in 1940 in St. Mary's, Pennsylvania, Engel received a diploma from Mercy Hospital School of Nursing, Buffalo, in 1961, then a bachelor's degree in public school nursing from Pennsylvania's Clarion University eight years later. After entering the Nurse Corps in 1969, Engel served in naval hospitals stateside and in Japan and Sardinia, holding progressively more responsible administrative posts. She was deputy director of the corps under Stratton.

These two leaders maintained that "Navy Nursing IS Nursing Excellence." They kept Navy nursing in the mainstream of American nursing, which now emphasized disease prevention and health promotion. Similarly, Navy nurses encouraged good health practices with such programs as smoking cessation, prenatal and infant clinics, and women's health clinics. Ambulatory care included home health visits and same-day surgery, while a holistic approach to pregnancy provided multidiscipline services. Teams of nurses staffed family practice clinics, and others conducted education programs. To assist with these changes, eighteen specialty advisers, a new means to address specifically matters in each subspecialty, became active.

During the early 1990s, Navy nurses in the United States continued to either attend civilian universities or serve at hospitals, clinics, branch clinics, recruiting commands, or Marine Corps stations. Abroad, they were in hospitals and clinics, on hospital ships, in aircraft carriers, and with fleet surgical teams of the Fleet Marine Force. In March 1994, twenty-three Navy nurses staffed the Fleet Hospital in Zagreb, Croatia, in support of United Nations peacekeeping forces. Two years later, five Navy nurses served in the amphibious assault ship *Wasp* (LHD 1) off the coast of Bosnia-Herzegovina. As always, nurses trained hospital corpsmen.

They continued their tradition of rendering humanitarian aid in sundry disasters such as Operation Fiery Vigil in July 1991, when nurses helped evacuate twenty-two hundred Air Force and naval personnel and their dependents from Subic Bay after Mt. Pinatubo erupted. On the other side of the world, Navy nurses at Guantanamo Bay gave medical care to hundreds of Haitian refugees. During the Los Angeles riots of April 1992, nurses at Naval Hospital, Long Beach, helped provide medical support for the ten thousand federal and military personnel activated in the area. In that same year, Navy nurses helped in the aftermath of

Typhoon Omar in the Pacific and of Hurricane Andrew in Florida. In 1994 nurses from Naval Hospital, Long Beach, volunteered to assist victims of an earthquake at Los Angeles.

Shared Experiences
Rank and Command
As the 1990s unfolded, more line and Nurse Corps officers rose to flag rank with commensurate command responsibilities. Rear Adm. Marsha J. Evans became superintendent of the Naval Postgraduate School at Monterey, and Rear Adm. Katherine L. Laughton took over the Naval Space Command at Dahlgren, Virginia. Rear Adm. Veronica Z. Froman became director for Manpower and Personnel, J-1, Joint Staff, while Rear Adm. Barbara E. McGann served as ACNP for Total Programming for Manpower at the Bureau of Naval Personnel. Patricia A. Tracey became a rear admiral in 1993 and commanded the Naval Training Center, Great Lakes. Three years later, she became the first woman promoted to vice admiral.

Simultaneously, directors of the Nurse Corps also held flag rank. Rear Adm. Mariann Stratton began serving as director as well as assistant chief for Personnel Management in 1991. Three years later, Rear Adm. Joan M. Engel took over these two commands. The first Reserve Nurse Corps member, Maryanne I. Ibach, gained flag rank in 1990, followed by Nancy A. Fackler.

More novel commands became available to women, as a few examples will illustrate. In late 1990, Lt. Cdr. Darlene Iskra assumed command of the rescue and salvage ship *Opportune* (ARS 41). The following year Lt. Cdr. Deborah Cernes became CO of the fleet oiler *Cimarron* (AO 177). Rear Adm. Louise C. Wilmot became the first woman to command a naval base when she took over at the Philadelphia Navy Base in 1993, and in 1994 Capt. Susan Brooker assumed command of Naval Readiness Command 22. The following year Capt. Linda V. Hutton became commander of the Naval Air Station, Key West.

In addition to the directors, other Navy nurses took on positions of more responsibility. Eight captains were selected as commanding officers in 1992. One, Capt. Barbara A. Mencik, became CO, Naval Hospital, Long Beach, and another, Capt. Ann Langley, served as CO, Naval School of Health Sciences, Bethesda. By 1995 four Navy nurses were commanding officers.

Pregnancy and Motherhood

Until the 1970s, pregnancy and dependent children usually meant an automatic end to a woman's naval career. As working mothers became the norm in the American labor force, the Navy relaxed its prohibitions during the 1980s, and by the 1990s, it had accepted pregnancy as a normal female condition. Even so, the Navy had to grapple with issues such as the effect on readiness, lost time, workload of shipmates, and hazards in the work environment. Secretary of the Navy John H. Dalton announced the Navy's new pregnancy policy in February 1995, which concluded that pregnancy and parenthood do not preclude a naval career and stipulated various medical and counseling options available.

Sexual Harassment and Tailhook

Although Navy women made progress professionally in the early 1990s, some things stayed the same. The longtime problem of sexual harassment exploded with a fury at the Tailhook Association convention in September 1991. The Navy had decreed "zero tolerance" for sexual harassment the previous year, but apparently this message had not reached naval and Marine aviators. Their annual professional meetings often degenerated into drunken brawls with attendant lewd behavior.

As they made inroads into naval aviation, female pilots began going to these affairs, including the 1991 convention, held at the Hilton Hotel in Las Vegas. After attending panels and symposia by day, many male pilots partied at hospitality suites where alcohol was plentiful. Soon

drunken men lined the hotel hallway outside the suites, groping and paw-
ing any women who passed through this "gauntlet." At least twenty-six
women, half of them naval officers, were molested in the gauntlet. Such
behavior might have been brushed aside except that one victim was Lt.
Paula Coughlin, a helicopter pilot and an admiral's aide. She reported the
frightening episode to the admiral, who took no action. Determined that
women should not suffer such indignities to protect their naval careers,
she next notified Vice Adm. Richard M. Dunleavy, assistant CNO for Air
Warfare, and her complaint soon reached the highest levels. By now the
press had picked up portions of the story.

Chief of Naval Operations Frank B. Kelso II launched a Naval Investigative Service investigation, and simultaneously the Navy inspector
general began another. Both groups reported in the spring of 1992; neither investigation resulted in charges against any of the seventy perpetra-
tors and identified only two suspects. Exasperated, Coughlin went public
with her story, and the Navy received a hailstorm of bad publicity. The
Department of Defense inspector general began still another investigation amid charges of cover-ups and scapegoating. Accepting full responsibility, Lawrence Garrett resigned as secretary of the Navy in June. The
summer of 1992 brought revocation of two nominations for high commands and postponing promotions for about five thousand officers until
they could show they had not been at the Tailhook convention. In those
same summer months, the Navy ordered all naval personnel and civilian
employees to attend a one-day conference about sexual harassment.

The DOD inspector general's reports, released in September 1992
and February 1993, were sharply critical of the unproductive investigations of the Naval Investigative Service and the Navy inspector general.
The admirals at the top of these offices were held responsible. The DOD
report raised the number of officers for possible discipline to 117 and the
number of women assaulted to 83.

Three aviators were eventually tried at Legal Services in Norfolk in
December 1993, but the charges against them were dismissed. The Marine

captain who had allegedly assaulted Coughlin was acquitted of criminal charges at Quantico, and eventually all charges against him were dropped. About fifty other officers received administrative penalties. No one was punished for sexual harassment or assault. Finally, CNO Kelso took early retirement in April 1994.

And what of Lieutenant Coughlin? She did not receive accolades for courageously turning the spotlight on blatant sexual harassment and molestation. Almost from the beginning, rumors vilifying her circulated in the Navy, and other naval aviators ostracized her. In February 1994 Coughlin submitted a letter of resignation stating that "the covert attacks on me . . . have stripped me of my ability to serve." She could later take comfort in the $6.7 million awarded her in punitive and compensatory damages by a federal jury in a lawsuit against the Las Vegas Hilton and Hilton Hotels.

The Tailhook scandal had rocked the entire naval establishment, turning searing attention to the persistent harassment problem, and everyone hoped Navy men had learned from the experience. Unfortunately, this was not the case. Glaring newspaper headlines revealed a continuing pattern of sexual misconduct.

In Orlando, for example, three instructors pushed an enlisted woman into a steaming shower, then beat and kicked her. In San Diego, ten instructors harassed communications trainees. The careers of two admirals ended because of adulterous relationships. Two officers—one at Norfolk, the other at Pearl Harbor—faced charges of sexual harassment. A drunken chief petty officer repeatedly groped an enlisted woman while on a cross-country commercial flight. In response, CNO Adm. Jeremy M. Boorda ordered more instruction for the entire Navy. Then the head of the U.S. Pacific Command uttered an impolitic remark, trivializing the rape of a twelve-year-old girl by a sailor and two marines. It ended that admiral's career. Nevertheless, sexual misconduct continued.

In spite of repeated incidents, Navy women reported in a survey that from 1988 to 1995 sexual harassment had fallen from 66 percent to 53 percent of all Navy women who had experienced or heard of such conduct during the past year. But the American public remained pessimistic and thought that harassment would always be present in the military.

Another unresolved problem was that of lesbians in the Navy. Homosexuals, both male and female, had been barred from military service, but newly elected President William Clinton urged lifting the ban. The "don't ask, don't tell" policy, implemented in 1994, permitted homosexuals to serve as long as they did not openly engage in homosexual activities or unless the military discovered their orientation. In operation, the policy has, in fact, led to the discharge of more gays than before. How the military will resolve this long-standing problem is unknown.

Over the years, women have made great strides in the U.S. Navy. The unthinkable—women's permanent military service—had become possible. Theirs is the story of uneven and often uncertain progress. Their advancement in the Navy generally mirrored the situation of women in American society; the avenues that opened for all women soon had parallels in the Navy.

The Navy's personnel requirements governed its acceptance of women. It needed clericals in World War I, hence the yeomen (F). It needed women to fill stateside desk jobs in World War II, hence the WAVES. Even after the 1948 Women's Armed Forces Integration Act made women a permanent part of the military, the Navy had little use for them. Few took women's service seriously, and women fought an uphill battle for two decades just to survive in the Navy.

The 1970s brought massive changes for Navy women. The end of the draft and the beginning of the all-volunteer force once again caused personnel shortages, and once again the Navy needed women. It might not want women, but it needed them. Then Z-116 opened the floodgates of increased professional opportunities and the real possibility of a viable career in the Navy. As women rose in rank and assumed command responsibilities, more and more jobs became available. Finally, the Persian Gulf War and the repeal of the combat exclusion law added impetus to career potential for women. Still, Navy women encountered resentment, opposition, harassment, and sexism. There was always a lingering doubt about whether they were wanted in the U.S. Navy.

On the other hand, there was never any doubt about the acceptance of Navy nurses. The Navy needed nurses to care for sailors and marines, and few questioned their value or usefulness. Navy men did not fear competition from these professionals. Perhaps such strong support fostered an atmosphere of security that kept the Nurse Corps outside the turmoil that often affected other Navy women.

Since the Sacred Twenty first donned the blue capes of Navy nurses in 1908, the corps has grown in numbers, composition, training, and expertise. It has kept abreast of currents in the wider field of American nursing and has, in fact, surpassed civilian nursing in such areas as college degree requirements for its members. In wartime and peacetime, Navy nurses have served at home and abroad, caring for naval personnel and their dependents and carrying out their vital function of training hospital corpsmen. The Navy Nurse Corps enters the new century as its first superintendent, Esther V. Hasson, had envisioned it: "a dignified and respected body."

The story of all Navy women is one of dedication and valor. They proudly chose to serve their country as part of the U.S. Navy and pursued their goal with dogged determination. They would not be denied such an honorable calling. As in the eighteenth and nineteenth centuries, women's familiarity with maritime matters and their nursing skills have made them essential to the naval service.

Tributes to their service have been many. One of the most touching came in 1996, when the Navy named the guided missile destroyer *Hopper* (DDG 70) in honor of Rear Adm. Grace Hopper, who had led the Navy into the computer age. And in the fall of 1997, a lasting monument to Navy woman, as well as to all 1.8 million women who have served in the military, was dedicated: the Women in Military Service for America Memorial at the entrance to Arlington National Cemetery, which stands as a fitting reminder of all women who have volunteered to defend American freedom.

INDEX

SERIES EDITOR

THOMAS J. CUTLER has been serving the U.S. Navy in various capacities for more than fifty years. The author of many articles and books, including several editions of *The Bluejacket's Manual* and *A Sailor's History of the U.S. Navy*, he is currently the director of professional publishing at the Naval Institute Press and Fleet Professor of Strategy and Policy with the Naval War College. He has received the William P. Clements Award for Excellence in Education as military teacher of the year at the U.S. Naval Academy, the Alfred Thayer Mahan Award for Naval Literature, the U.S. Maritime Literature Award, and the Naval Institute Press Author of the Year Award.

The **Naval Institute Press** is the book-publishing arm of the U.S. Naval Institute, a private, nonprofit, membership society for sea service professionals and others who share an interest in naval and maritime affairs. Established in 1873 at the U.S. Naval Academy in Annapolis, Maryland, where its offices remain today, the Naval Institute has members worldwide.

Members of the Naval Institute support the education programs of the society and receive the influential monthly magazine *Proceedings* or the colorful bimonthly magazine *Naval History* and discounts on fine nautical prints and on ship and aircraft photos. They also have access to the transcripts of the Institute's Oral History Program and get discounted admission to any of the Institute-sponsored seminars offered around the country.

The Naval Institute's book-publishing program, begun in 1898 with basic guides to naval practices, has broadened its scope to include books of more general interest. Now the Naval Institute Press publishes about seventy titles each year, ranging from how-to books on boating and navigation to battle histories, biographies, ship and aircraft guides, and novels. Institute members receive significant discounts on the Press' more than eight hundred books in print.

Full-time students are eligible for special half-price membership rates. Life memberships are also available.

For a free catalog describing Naval Institute Press books currently available, and for further information about joining the U.S. Naval Institute, please write to:

Member Services
U.S. NAVAL INSTITUTE
291 Wood Road
Annapolis, MD 21402-5034
Telephone: (800) 233-8764
Fax: (410) 571-1703
Web address: www.usni.org